02915H

D0310548

INVENTION & DISCOVERY

STRUAN REID

Designed by Sue Mims

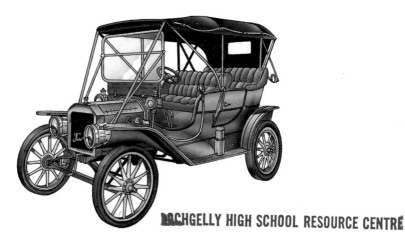

LOCHGELLY HIGH SCHOOL RESOURCE CENTRE

Consultants

Brian Adams, Roger Price, Tom Petersen, Sue Watt, Stephen Brooks, John and Sue Becklake

Illustrations by

Guy Smith, Chris Lyon, Martin Newton, Richard Draper

Additional illustrations by

Jeremy Gower, Kevin Maddison, Gerard Browne, Ian Jackson, Paul Cooper

Contents

First published in 1986
Usborne Publishing Ltd.,
20 Garrick Street,
London WC2E 9BJ, England.

Copyright ©1986 Usborne Publishing Ltd.

The name Usborne and the device 🐝 are Trade Marks of Usborne Publishing Ltd.

About this book

This illustrated history describes the most important inventions and discoveries that have marked the progress of the human race, from the Stone Age to the Space Age. It tells you when something was invented or discovered, what important improvements were made and what the latest version is.

The book is arranged thematically under general topics. Each topic is grouped in sections, identified by a coloured triangle at the bottom right-hand side of each double page, as described below:

- ◢ Energy and power
- ◢ Tools, machines and building
- ◢ Transport
- ◢ Farming and food
- ◢ Communication
- ◢ Domestic appliances

- ◢ Scientific instruments
- ◢ Medicine, chemistry and biology
- ◢ Warfare
- ◢ Chronological list of inventions and discoveries
- ◢ A-Z of inventors

Using this book

Each entry has been arranged with the date of the invention or discovery, the name of the person responsible, the country he or she came from and the invention or discovery itself. In the case of the earliest dates, it is not always possible to be sure exactly when the invention or discovery was made. In these cases, the earliest known examples are referred to.

The country in brackets after the person's name is the modern name for the country he or she came from. There is one exception to this rule - Mesopotamia. This means 'the land between the two rivers', the area between the Tigris and Euphrates rivers in the Middle East. This name for the area has been given as it does not correspond exactly to any country today. If there are two country names, it means that the person was born in one country but moved to another. Abbreviations are used for the following countries:

Great Britain, 'GB'
New Zealand, 'NZ'
United States of America, 'USA'
Union of Soviet Socialist Republics, 'USSR'.

In front of some of the dates, particularly the early ones, you will find the letter 'c', like this: c.2000BC. This stands for the Latin word 'circa', meaning 'about'. This means that the dates given are approximate.

Many inventions have a 'patent'. This is a document giving the inventor the sole right to make and sell the invention for a set period. The patent date is often the only known date of an invention, despite the fact that it may have existed some time before being patented.

If a word is in bold type and has an asterisk after it, like this: **Steam engine***, it is in a footnote at the bottom of the page and refers you to the main entry elsewhere in the book.

Fuels and mining equipment

Wood was the first fuel used for heating and cooking. By **5000BC**, **charcoal** was used for smelting metals in Europe and the Middle East. It produced a much stronger heat than wood. The most common fuels used today are coal and oil. They are known as 'fossil fuels' as they were formed from plants and trees (coal) and sea creatures (oil) that lived millions of years ago.

Coal

- **c.1st century AD** The **Romans** first used **coal** as a fuel in northern Europe.

- **16th century** Coal became a **common fuel** in Britain. It was dug out of the ground with pick axes and carried to the surface in carts.

- **1709 Abraham Darby** (GB) produced **coke** to use in **blast furnaces***. It was made by partly burning coal in a closed chamber. This produced a high-carbon fuel which was cleaner and hotter than coal.

- **1760 W.Brown** (GB) invented the first **coal-cutting machine**. It was powered by horse and aimed a pick at the coal face, hitting it harder and more often than a man could.

◄ **Cutaway diagram of Davy's safety lamp**

The flame of an oil lamp was enclosed by wire mesh. The mesh absorbed the heat of the flame before it could come in contact with the gas in the air and cause an explosion.

- **1815 Humphry Davy** (GB) designed a **miner's safety lamp**, to reduce the risk of explosions of methane gas in pits.

Battery*-powered lights ◄ are now fitted to miner's helmets.

- **1830 Lord Cochrane** (GB) designed a **compressed-air engine**, which pumped air down to machines in the mines, providing power to cut the coal. **Steam engines*** would have made too much smoke.

- **1863 T.Harrison** (GB) designed a **cutter with sharp blades** driven by air.

- **1910** The first colliery to have **all electric mining equipment** was opened at Monmouth, Wales.

- **1946 Dowty Co.** (GB) introduced the first **hydraulic pit props**. Pressure is transmitted through a liquid to pump up props that support the roof of the pit.

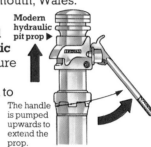

Modern hydraulic pit prop ►

The handle is pumped upwards to extend the prop.

- **1963** A system called **Remotely Operated Longwall Face (ROLF)** (GB) was introduced. An engineer operates a remote-controlled **cutting machine** from an underground control panel. As the face is cut back, **hydraulic props** move forward to hold up the roof of the pit.

Cutaway diagram of a modern coal mine ▼

Office where engineers monitor the machines underground and on the surface.

6.Coal skip is loaded from the hopper and raised to the pithead. From here the coal is taken on a conveyor belt to be washed and prepared for burning.

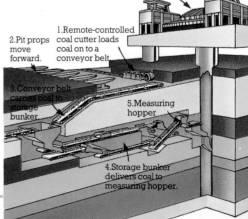

1.Remote-controlled coal cutter loads coal on to a conveyor belt.

2.Pit props move forward.

3.Conveyor belt carries coal to storage bunker.

4.Storage bunker delivers coal to measuring hopper.

5.Measuring hopper

4

Oil

- **c.2400BC Bitumen**, a form of oil that seeps from the ground, was used in Mesopotamia to make boats watertight.

- **AD1859 E.Drake** (USA) drilled the first **oil well** at Titusville, USA. He found oil 21m (69½ft) below the surface. The method he used is called **cable-tool drilling**. A metal tool on the end of a cable was dropped on to the rock, smashing it up.

Wooden tower containing the drill.

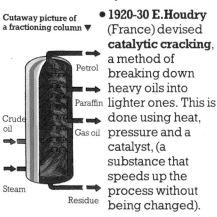

Drake's drilling rig was made of wood. ▶

- **1884 R.Beart** (GB) introduced **rotary drilling** powered by a **steam engine***. A drilling bit was mounted on the end of a hollow steel pipe. The rock cuttings were forced up by pumping water down the hollow drill pipe.

- **c.1900** The first **offshore oil well** was built off the coast of California, USA.

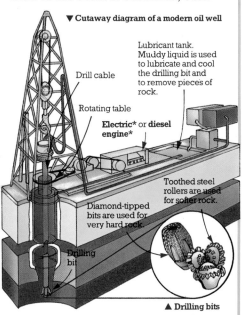

▼ Cutaway diagram of a modern oil well

Lubricant tank. Muddy liquid is used to lubricate and cool the drilling bit and to remove pieces of rock.

Drill cable

Rotating table

Electric* or **diesel engine***

Toothed steel rollers are used for softer rock.

Diamond-tipped bits are used for very hard rock.

Drilling bit

▲ Drilling bits

Oil refining

- **1850 J.Young** (GB) started the commercial production of **paraffin** from crude oil made from heated coal. The crude oil was distilled into its components, or fractions, in containers heated by steam.

Cutaway picture of a fractioning column ▼

Petrol

Paraffin

Crude oil

Gas oil

Steam

Residue

- **1920-30 E.Houdry** (France) devised **catalytic cracking**, a method of breaking down heavy oils into lighter ones. This is done using heat, pressure and a catalyst, (a substance that speeds up the process without being changed).

Steam boils the oil which vaporizes. The vapour collects at different heights according to its contents.

Gas

- **c.1791 P.Lebon** (France) obtained **gas by heating charcoal** and began experimenting with gas for lighting.

- **1792 W.Murdock** (GB) first used **coal gas lighting** in a house in Redruth, Cornwall, England.

- **1980's Natural gas** is the most common gas used today. It is found in pockets under the sea bed.

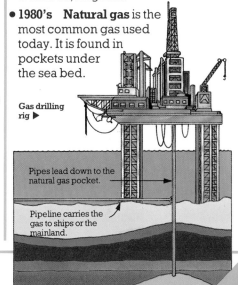

Gas drilling rig ▶

Pipes lead down to the natural gas pocket.

Pipeline carries the gas to ships or the mainland.

*** Battery**, 6; **Blast furnace**, 22; **Diesel engine**, 9; **Electric engine**, see **Motors**, 7; **Steam engine**, 8.

Electricity, magnetism and motors

Electricity is a form of energy made by the movement of tiny particles called **electrons***. The movement of these electrons is called an electric current. Natural electricity can be seen in the form of lightning. In animals, a beating heart generates a form of electricity. But most electricity is man-made, generated in power stations for lighting, heating and other purposes.

How a simple wet-cell ▶
battery works

1.Cell containing a liquid called an electrolyte. It is made of billions of particles with **positive** and **negative charges***.

2.Electrodes (rods of zinc and copper) are immersed in the electrolyte. A chemical reaction in the electrolyte sends positive particles to one electrode and negative particles to the other.

3.A conductor (wire) is connected to the electrodes. The chemical reaction produces an electric current which flows through the wire.

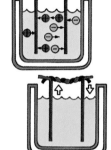

When an electric current flows through a conductor it generates heat. With a strong current, the conductor will glow red. This principle is used to make **electric light bulbs***. ▼

Electricity and batteries

When lightning ▲ struck the **lightning conductor** it took the quickest route - down the conductor instead of the house.

● **1752 Benjamin Franklin** (USA) proved the existence of **static** (natural) **electricity**. Using a metal key attached to the string of a kite, he made a flash of lightning travel down the string to the ground. The same year he installed the first **lightning conductor**.

● 1780 **L.Galvini** (Italy) first demonstrated **animal electricity**. The twitching effect produced by an electric current on a pair of frog's legs showed the connection between muscle activity and electricity.

● 1800 **Alessandro Volta** (Italy) made the first **wet-cell battery** to produce an **electric current**. He discovered that when zinc and silver discs (or electrodes) were placed in an acid solution, an electric current flowed through a wire linking the discs.

Electrons bounce off the **atoms*** in the conductor. This makes the atoms vibrate, producing heat and light.

● **1839 W.Grove** (GB) made the first **fuel cell**. This produced electricity from a chemical reaction between oxygen and hydrogen in a container.

● 1866 **G.Leclanche** (France) made the first **dry-cell battery**.

Metal case

Cutaway picture of a modern
dry-cell battery (used in torches and radios). ▶

Metal cap carries electric current.

Electrolyte layer kept dry in paper.

Contents of manganese dioxide, carbon, ammonium chloride and zinc chloride.

The electric current is made by a reaction between the zinc cup, chemical contents and electrolyte.

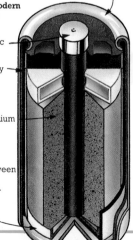

Zinc cup

Volta's
battery ▶

Zinc and silver discs separated by damp fabric pads

Acid solution

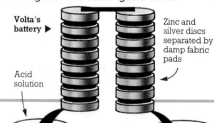

Electricity and magnetism

• **1820 H.Oersted** (Denmark) discovered the connection between **electricity and magnetism**. He noticed that an electric current in a wire attracted a **compass*** needle nearby. This is because whenever an electric current flows it creates a **magnetic field**.

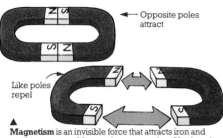

← Opposite poles attract

Like poles repel

▲ **Magnetism** is an invisible force that attracts iron and steel to a magnet. Magnets have two poles - North and South. When opposite poles of two iron magnets are put next to each other, they attract one another. Two north poles or two south poles face to face repel each other.

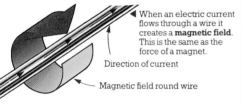

◄ When an electric current flows through a wire it creates a **magnetic field**. This is the same as the force of a magnet.

Direction of current

Magnetic field round wire

• **1823 W.Sturgeon** (GB) made the first **electromagnet**. By passing an electric current through a wire wrapped round an iron bar he created a magnet which could lift 20 times its own weight.

How an electromagnet works ▼   Wire wrapped round iron bar

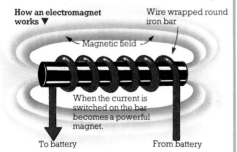

Magnetic field

When the current is switched on the bar becomes a powerful magnet.

To battery   From battery

• **1831 J.Henry** (USA) made **more powerful electromagnets** by insulating the wire with silk. Using a small battery, Henry made an electromagnet lift weights up to 946kg (2,086lb).

Electric motors, transformers and dynamos

• **1821 M. Faraday** (GB) made the first basic **electric motor**. He placed a current-carrying wire between the poles of a magnet. The two magnetic fields made the wire rotate.

Faraday's electric motor ▼

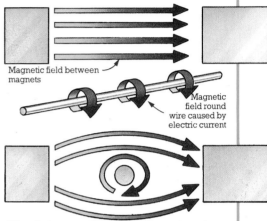

Magnetic field between magnets

Magnetic field round wire caused by electric current

When the two magnetic fields meet they produce a force which turns the wire round.

• **1831 M. Faraday** (GB) invented the **transformer**, a device for changing the strength of an electric current (voltage).

How a transformer ▼ **works**

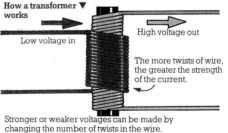

Low voltage in   High voltage out

The more twists of wire, the greater the strength of the current.

Stronger or weaker voltages can be made by changing the number of twists in the wire.

• **1832 H.Pixii** (France) made the first **dynamo**, a device for making electricity. When an electric conductor is placed near a magnet, an electric current is created in the conductor. This action, combined with magnetism produces electricity.

• **1899 N.Tesla** (Yugoslavia/USA) designed the first **small motor** which could be fitted to **domestic appliances***.

---

* **Atoms**, 14; **Compass**, 72; **Domestic appliances**, 64; **Electric light bulb**, 62; **Electrons**, 14; **Negative charge**, **Positive charge**, 14.

Engines

An engine is designed to run vehicles and machines. There are many different kinds of engine.

The earliest was the steam engine, whose invention paved the way for the Industrial Revolution in Europe.

Steam engines

- **1698 T.Savery** (GB) patented the first **steam engine**. It was used to pump water from flooded mines. Steam from a boiler passed into a cylinder where it was condensed back into water again. This created a lower pressure inside the cylinder which sucked up the water.

- **1712 Thomas Newcomen** (GB) fitted a **piston** to his steam engine, which was used to drive water pumps.

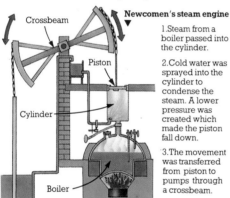

Newcomen's steam engine ▼

Crossbeam

Piston

Cylinder

Boiler

1. Steam from a boiler passed into the cylinder.

2. Cold water was sprayed into the cylinder to condense the steam. A lower pressure was created which made the piston fall down.

3. The movement was transferred from piston to pumps through a crossbeam.

- **1769 James Watt** (GB) added a **separate condenser** where steam from the engine cylinder was condensed. The engine could now be kept hot instead of having to cool down to condense the steam before being reheated. This reduced fuel consumption and saved time.

- **1782 James Watt** (GB) introduced **sun and planet gears**, which converted the up-and-down movement of the engine to **rotary motion**.

Rotary motion ▶

The planet gear was fixed at the end of a connecting rod and turned the sun gear which was fixed to the flywheel.

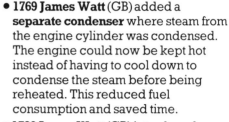

Flywheel

Belt on flywheel took the power to factory machines.

Sun gear

- **c.1800 R.Trevithick** (GB) built a smaller, **high-pressure engine**. It produced about 5 times more pressure than Watt's engine.

Internal combustion engines

- **1860 E.Lenoir** (Belgium) designed the first **internal combustion engine**. It was based on the steam engine but used **gas*** instead. The gas was ignited inside a cylinder by sparks. This explosion pushed forward a piston inside the cylinder. The piston was attached to a flywheel on a crank which could drive machinery.

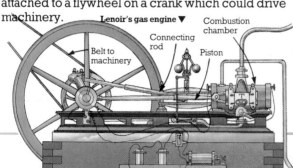

Lenoir's gas engine ▼

Belt to machinery

Connecting rod

Combustion chamber

Piston

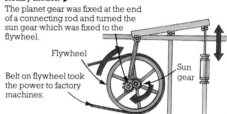

- **1862 A.de Rochas** (France) patented the first **four-stroke engine design**, using a **petrol***-air mixture. This was the basis of the modern internal combustion engine.

- **1876 N.Otto** (Germany) built the first **gas four-stroke engine**. It was four times more efficient than Lenoir's engine.

- **1883 G.Daimler** (Germany) built the first **petrol engine**.

Diagram of one cylinder from a **modern four-stroke car engine**. Four stages are needed to get the engine moving. These are done in quick succession.

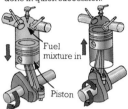

Fuel mixture in

Piston

1. Descending piston draws petrol-air mixture into cylinder.
2. Piston moves up and compresses mixture.
3. Mixture is ignited by spark, explodes and drives the piston down.
4. Piston moves up again and forces out burnt gases through exhaust valve.

Piston connected to a crankshaft which rotates and starts the vehicle moving.

- **1890 H.Stuart** (GB) designed an engine called a **compression-ignition engine**. It was given this name because the fuel was not detonated by a spark, but by air heated by compression in the cylinder.

- **1892 R.Diesel** (Germany) improved the design of the compression-ignition engine, which has been known since as the **diesel engine**. It is simpler than other engines because it needs no spark-plugs.

Rotary engine

The rotary engine ▶

- **Late 1950s F.Wankel** (Germany) designed the **rotary engine**. This has no pistons and rods and moves more smoothly than other engines.

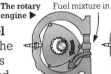

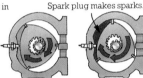

Fuel mixture in

Spark plug makes sparks.

Burnt gases out

1. Turning rotor draws fuel mixture into chamber.
2. Mixture compressed into smaller space by turning rotor.
3. Mixture explodes and drives rotor round.
4. As rotor turns, burnt gases escape through exhaust hole.

Turbine engines

- **1884 C.Parsons** (GB) built the first practical **steam turbine***. Steam turbines are still used to generate **electricity***.

Steam turbine ▼

Steam hits next set of vanes which are attached to a rotating shaft.

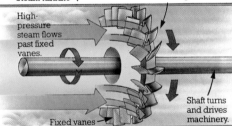

High-pressure steam flows past fixed vanes.

Fixed vanes

Shaft turns and drives machinery.

- **1936 Brown Boveri Co.** (Switzerland) installed the first **gas turbine** at an oil refinery in the USA. In a gas turbine, air is drawn in by a fan and compressed. Fuel is then mixed with the compressed air and burns in the combustion chamber. The gases turn turbine blades.

Jet engine

- **1930 Frank Whittle** (GB) patented his idea of the **jet engine** for aircraft. It was based on the gas turbine. The first **jet aeroplane*** flew in 1939.

How a modern turbofan* jet engine works ▼

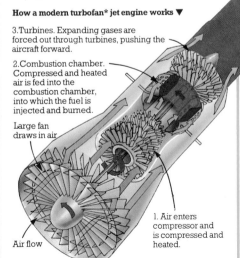

3. Turbines. Expanding gases are forced out through turbines, pushing the aircraft forward.

2. Combustion chamber. Compressed and heated air is fed into the combustion chamber, into which the fuel is injected and burned.

Large fan draws in air

Air flow

1. Air enters compressor and is compressed and heated.

Electricity, 6; *Gas*, 5; *Jet aeroplane*, 35; *Petrol*, 5; *Steam turbine*, see also *Water turbines*, 10; *Turbofan*, 118.

9

Water power and wind power

Before the invention of the **steam engine***, power was provided by wind, water and animals. The earliest references to water mills appeared in Greek writings of the 1st century BC. Due to the expense of traditional fuels and the fear that they might run out, people have been looking at alternative sources of power, many of which have very ancient histories.

Water wheels

- By **70BC** the **Romans** had developed two types of wooden **water-wheel** to grind grain and olives.

The undershot wheel ▶

The lower paddles were moved round by the flowing stream.

The overshot wheel ▶

The water was directed on to the top of the wheel, pushing the wheel round.

Water turbines

- **1827 B.Fourneyron** (France) designed the first practical **water turbine**, a machine that uses energy from moving water to drive machinery. By the end of the 19th century, turbines had replaced water wheels in factories.

- **1850s J.Francis** (USA) designed the **reaction turbine**, which used the water more efficiently than the earlier Fourneyron model.

3.The pressure of water on the shaft vanes rotates the shaft, which in turn drives the machinery.

2.Guide vanes direct the flow of water into the centre of the turbine.

1.Water flows into the turbine through a narrow tube. This increases the speed of the water.

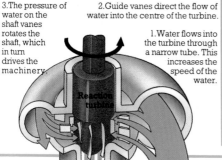

Reaction turbine

- **1880s L.Pelton** (USA) designed the **impulse turbine**.

Impulse turbine ▶

Jet of water directed on series of buckets mounted round rim of wheel, driving it round.

The turbine can be stopped by deflecting the jet of water away from the buckets.

- **1910-24 V.Kaplan** (Germany) designed the **axial flow turbine**.

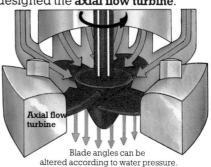

Axial flow turbine

Blade angles can be altered according to water pressure.

Hydroelectric power (HEP) stations

- **1891** The first **hydroelectric power stations** were built near Frankfurt, Germany and at the Niagara Falls, USA. Hydroelectric power is produced by water-driven turbines and generates a quarter of the world's **electricity***.

Diagram of an HEP station ▼

When the sluice gate is raised, water held in a reservoir behind the dam rushes at high pressure along a tunnel.

Sluice gate

In the powerhouse, water turns the turbine and a generator attached produces electricity.

Dam and powerhouse

Electricity is transmitted by cables.

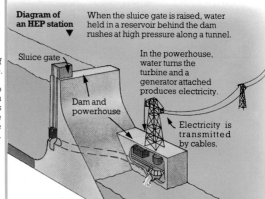

Tidal power

- **1086 The Domesday Book** (England) described a **tidal mill** at Dover. Tides have been used to provide power for water wheels for at least 900 years.

- **1968** The first **tidal power station**, using tides to power water turbines, was built across the River Rance, France.

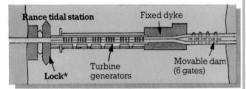

Rance tidal station | Fixed dyke
Turbine generators | Movable dam (6 gates)
Lock*

A 750m (2,460ft) long dam traps water from the sea at high tide. It is then released through 24 turbines which generate **electricity***.

Wave power

- **1974 S.Salter** (GB) designed a **wave-power machine** for generating **electricity***.

A series of floats were mounted on a concrete spine. Each float had a pointed end facing into the oncoming waves and they rocked up and down.

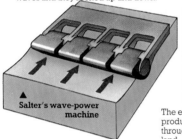

▲ Salter's wave-power machine

Nodding motion of the float made **gyroscopes*** inside rotate. This action pumped hydraulic fluid to a generator.

The electricity produced travelled through a cable to land.

- **1965 Y.Masuda** (Japan) used the principle of the **oscillating** (moving) **water column** to generate **electricity***.

A large concrete box, resting on the sea bed, had its front open to waves.

Incoming wave raised the level of the water in the box, so compressing the air above it. The air was forced through an air turbine which drove a generator.

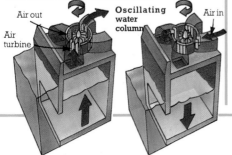

Air out
Air turbine
Oscillating water column
Air in

Windmills

- By **644** the **Persians** were using **windmills** with **vertical axes**.

▲ Vertical windmill

- **1180** The first reference to a **European windmill** appeared in France. By this time **horizontal sails** were used.

▲ Horizontal windmill

- **1920s M.Jacobs** (USA) designed a windmill to work as an **electricity*** **generator**. He used a three-bladed propeller which spun faster.

- **1920s G.Darrieus** (France) designed a wind generator with a **vertical axis**.

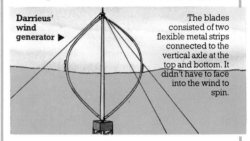

Darrieus' wind generator ▶

The blades consisted of two flexible metal strips connected to the vertical axle at the top and bottom. It didn't have to face into the wind to spin.

- **1941 P.Putnam** (USA) completed a **giant wind generator** in Vermont, USA.

◀ **Putnam's wind generator** consisted of a 33.5m (110ft) high tower with a two-bladed propeller.

- **1970s P.Musgrove** (GB) designed a **hinged-blade generator**, with variable speed control.

Musgrove's generator had ▶ a vertical axis with 3m (10ft) hinged blades fixed to a horizontal beam. If the wind was too strong the blades folded outwards, which reduced the speed.

Electricity, 6; Gyroscope, 20; Lock, 27; Steam engine, 8.

Solar energy

The potential of solar energy has been understood for hundreds of years. In the 5th century BC, the Greek philosopher Socrates recommended that houses should be built in positions that would make the most of the Sun.

- 1861 **A.Mouchet** (France) patented the first **solar engine**. A mirror focused the Sun's rays on to a small water boiler which produced steam and drove a **steam engine***.

- 1872 **C.Wilson** (GB) designed the first **solar distillation system** at Las Salinas, Chile. The heat of the Sun was used to evaporate salty water, leaving the salt behind.

Solar heating

The commonest use of solar power today is for **heating in houses** and **waterheating**, using flat collectors.

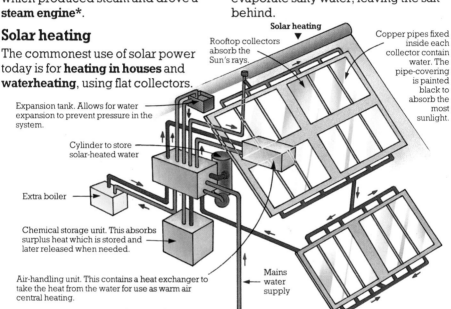

Solar heating ▼

Rooftop collectors absorb the Sun's rays.

Copper pipes fixed inside each collector contain water. The pipe-covering is painted black to absorb the most sunlight.

Expansion tank. Allows for water expansion to prevent pressure in the system.

Cylinder to store solar-heated water

Extra boiler

Chemical storage unit. This absorbs surplus heat which is stored and later released when needed.

Air-handling unit. This contains a heat exchanger to take the heat from the water for use as warm air central heating.

Mains water supply

Solar electricity

- 1954 **Bell Telephone Laboratories** (USA) first used **solar cells** to produce a device for generating **electricity***. The cells generate electricity from the Sun's rays.

Photon, a unit of light, coming from the Sun.

Current collector. Electrons are attracted to this and enter the electric circuit.

Silicon and boron layer

Silicon layer

Electron

◀ A **solar cell** consists of a thin layer of silicon next to another layer of silicon mixed with boron. Light falling on the outer silicon layer causes **electrons*** to move into the other layer, creating an electrical charge between the two layers.

- 1968 **P.Glaser** (USA) suggested placing an array of **solar cells in orbit**, 37,013km (23,000 miles) above the Equator, for generating electricity and sending it to Earth.

The photon dislodges one of the electrons, causing an electrical charge between the two layers.

Ocean thermalelectric conversion (OTEC) systems

- **1881** J. d'**Arsonval** (France) first suggested using **solar heat stored in the sea** to drive an engine.
- **1980s OTEC station** was started off the coast of Florida, USA to **collect solar energy from the sea**.

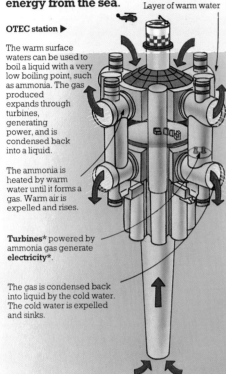

Layer of warm water

OTEC station ▶

The warm surface waters can be used to boil a liquid with a very low boiling point, such as ammonia. The gas produced expands through turbines, generating power, and is condensed back into a liquid.

The ammonia is heated by warm water until it forms a gas. Warm air is expelled and rises.

Turbines* powered by ammonia gas generate **electricity***.

The gas is condensed back into liquid by the cold water. The cold water is expelled and sinks.

Solar furnaces

- **1969** The first successful **solar furnace** was built at Ordeillo, France. It produces steam to generate **electricity***.

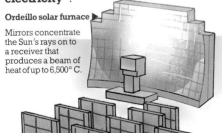

Ordeillo solar furnace ▶

Mirrors concentrate the Sun's rays on to a receiver that produces a beam of heat of up to 6,500° C.

Geothermal steam

- **1904** The first **geothermal power station** was built in Tuscany, Italy. Geothermal energy comes from the heat stored in the Earth's core.

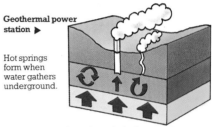

Geothermal power station ▶

Hot springs form when water gathers underground.

Steam emerges through cracks in the ground at temperatures of up to 260°C. This steam can be fed directly to **steam turbines*** to generate **electricity***.

Hot rocks

- **1974** The first **pilot holes** were bored in New Mexico, USA, to reach hot rocks where the steam does not emerge naturally.

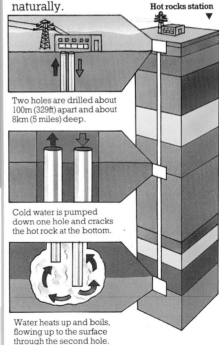

Hot rocks station ▼

Two holes are drilled about 100m (329ft) apart and about 8km (5 miles) deep.

Cold water is pumped down one hole and cracks the hot rock at the bottom.

Water heats up and boils, flowing up to the surface through the second hole.

At the surface the hot water passes through a heat exchanger, transferring the heat to air blown over it.

** **Electricity**, 6; **Electron**, 14; **Steam engine**, 8; **Steam turbine**, 9.*

Nuclear physics

Atomic structure

- **1808 J.Dalton** (GB) published his **atomic theory**. It stated that all **matter*** was composed of extremely small particles, called atoms.

▲ Dalton's idea of the atom consisted of a solid ball.

- **1897 J.Thomson** (GB) discovered **electrons**, particles in an atom with negative electric charge.

- **1907 J.Thomson** (GB) proposed that the atom consisted of a positively-charged sphere with electrons distributed through it. This was known as the **"plum pudding" model**.

▲ Thomson's "plum pudding" model of the atom

- **1911 E.Rutherford** (NZ) proposed the **nuclear model of the atom**. In this, the atom's positive charge and most of its **mass*** are concentrated in a tiny nucleus (core) about 10,000 times smaller than the atom. The electrons orbit the nucleus.

Rutherford's nuclear model of the hydrogen atom ▼

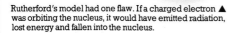

Nucleus · Electron · Electron · Nucleus

Rutherford's model had one flaw. If a charged electron ▲ was orbiting the nucleus, it would have emitted radiation, lost energy and fallen into the nucleus.

- **1913 N.Bohr** (Denmark) modified Rutherford's model to incorporate **quantum ideas***, that energy comes in fixed units, called 'quanta'.

Bohr's model explained ▶ why electrons did not fall into the nucleus. He proposed that orbiting electrons could only exist in well-defined, fixed energy (quantized) orbits. In these orbits they would not emit energy and could therefore continue orbiting indefinitely. Radiation is only emitted if an electron moves from one orbit to another.

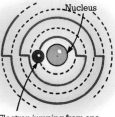

Nucleus

Electron jumping from one fixed orbit to another

- **1919 E.Rutherford** (NZ) identified **protons**, particles in an atom with positive electric charge.

- **1932 J.Chadwick** (GB) discovered **neutrons**, particles in an atom with no electric charge.

Model of the atom according to the latest ideas ▼

The nucleus of an atom consists of protons and neutrons. Electrons orbit the nucleus in definite quantized shells (orbits).

Protons. These have positive electric charge. In an electrically neutral atom, the number of protons in the nucleus = the number of electrons orbiting the nucleus.

Neutrons. These have no electric charge.

Electrons. These have negative electric charge.

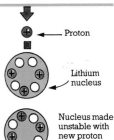

Particle accelerators

- **1930 J.Cockcroft** (GB) and **E.Walton** (Ireland) built the first atomic particle accelerator known as an **electrostatic accelerator**. But the insulation in this machine was poor at very high voltages. A particle accelerator is used to shatter atomic particles and provide information on their structure. Hundreds of new atomic particles have since been discovered, such as hyperons, pi-mesons, k-mesons and neutrinos.

Cockroft and Walton's experiment ▶

A very high voltage was used to accelerate protons through a **vacuum*** in a tube. These protons were fired at a target made of a metal called lithium. The lithium nuclei that were hit by the protons flew apart, changing into a gas called helium.

Proton

Lithium nucleus

Nucleus made unstable with new proton

Helium nucleus

Helium nucleus

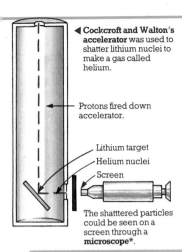

◀ **Cockcroft and Walton's accelerator** was used to shatter lithium nuclei to make a gas called helium.

Protons fired down accelerator.

Lithium target

Helium nuclei

Screen

The shattered particles could be seen on a screen through a **microscope***.

● **1931 E.Lawrence** and **M.Livingston** (USA) built a type of accelerator called a **cyclotron**.

Lawrence and Livingston's cyclotron ▶

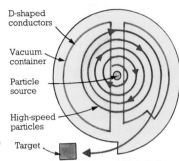

D-shaped conductors

Vacuum container

Particle source

High-speed particles

Target

At the heart of the cyclotron were two D-shaped conductors, between which there was an **alternating voltage***. Inside the D-shaped conductors there was a **magnetic field** which bent the particles into a circular path. When the particles crossed from one conductor to another they were accelerated by the voltage in a spiral orbit to the target.

The cyclotron has developed into the modern **synchrotron**. This is similar to the cyclotron except that the accelerated particles do not have a spiral orbit. Instead they are kept in a constant orbit by increasing the magnetic field.

Radioactivity

● **1896 A.Becquerel** (France) discovered **radioactivity** in a metal called **uranium***. Radioactivity is the spontaneous disintegration of the nucleus of an atom, giving off radiation.

● **1898 Pierre** and **Marie Curie** (France) discovered the radioactive **elements*** **radium** and **polonium**. By isolating these, they established that radioactivity is a property of a particular atom and that only certain elements produce radioactive atoms.

● **1899 E.Rutherford** (NZ) distinguished two types of radiation emitted by uranium: **alpha particles** (low penetration radiation, easily stopped) and **beta particles** (a more penetrating form of radiation).

● **1905 Albert Einstein** (Germany) worked out the formula $E=mc^2$, which showed that mass can be converted into energy. This knowledge led to the development of the **atomic bomb***.

E=mc²

Einstein's equation states that the energy (E) contained in any particle of **matter*** is equal to the **mass*** (m) of the matter multiplied by the square of the speed of light (c²), which is 299,330km/186,000 miles per second. A tiny amount of matter can therefore release huge amounts of energy.

● **1906 P.Villard** (France) showed that radium emits a third type of radiation called **gamma rays**. These have even greater penetration and need several metres of lead to absorb them. They have properties similar to **x-rays*** but have shorter **wavelengths***.

Radioactivity ▼

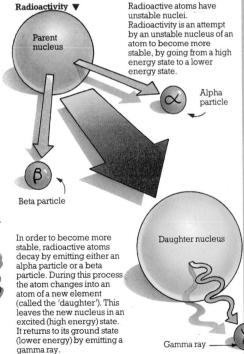

Parent nucleus

Alpha particle

Beta particle

Daughter nucleus

Gamma ray

Radioactive atoms have unstable nuclei. Radioactivity is an attempt by an unstable nucleus of an atom to become more stable, by going from a high energy state to a lower energy state.

In order to become more stable, radioactive atoms decay by emitting either an alpha particle or a beta particle. During this process the atom changes into an atom of a new element (called the 'daughter'). This leaves the new nucleus in an excited (high energy) state. It returns to its ground state (lower energy) by emitting a gamma ray.

** **Alternating voltage**, 118; **Atomic bomb**, 92; **Elements**, 88; **Magnetic field**, 7; **Mass**, **Matter**, 118; **Microscope**, 77;*
*__Quantum ideas__, see **Max Planck**, 116; **Uranium**, 16; **Vacuum**, 118; **Wavelength**, see **Laser**, 21; **X-rays**, 85.*

15

Nuclear power

Nuclear power is the power released by the atomic nucleus. When the nucleus (core) of an atom divides into two, in a process called 'fission', it releases huge amounts of energy. Released very quickly, this will produce a **nuclear bomb***, but done slowly and under control it can be used to generate **electricity***. The type of atom used is called uranium-235, an **isotope*** of the metal uranium.

Nuclear fission

● **1939 O.Hahn** (Germany) announced his discovery of **nuclear fission**. When bombarded by neutrons, the nuclei of uranium-235 atoms break up, releasing great amounts of energy.

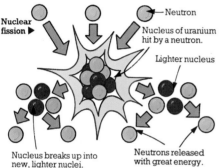

Nuclear fission ▶

Neutron

Nucleus of uranium hit by a neutron.

Lighter nucleus

Nucleus breaks up into new, lighter nuclei.

Neutrons released with great energy.

● **1942 E.Fermi** (Italy) assembled the first **nuclear reactor** in Chicago, USA. This produced energy by fission. Neutrons produced by fission normally travel very fast. If they travel too fast, they cannot split open other nuclei and create more energy. However, a fast neutron can be slowed down by making it bounce off a lighter material known as a **moderator** and so increase its chances of fission.

Moderator atoms slow down the neutrons.

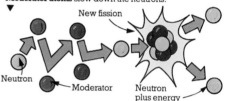

New fission

Neutron

Moderator

Neutron plus energy

● **1957** The first **pressurized water reactor** (PWR) was built in Pennsylvania, USA. Many modern power stations are based on this design.

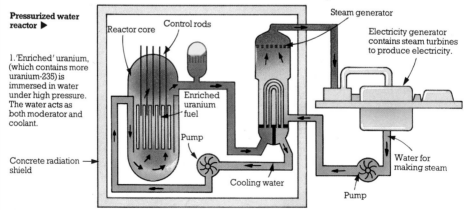

Pressurized water reactor ▶

Reactor core

Control rods

Steam generator

Electricity generator contains steam turbines to produce electricity.

1.'Enriched' uranium, (which contains more uranium-235) is immersed in water under high pressure. The water acts as both moderator and coolant.

Enriched uranium fuel

Pump

Concrete radiation shield

Cooling water

Water for making steam

Pump

2.The water is heated by the reactor core and pumped to the steam generator where it boils more water to form steam.

3.The steam is used to drive **steam turbines*** in the electricity generator.

Cutaway diagram of Fermi's reactor. It was made up of blocks of a type of carbon called graphite to act as the moderator. Running through the pile were rods of cadmium, a metal that absorbs neutrons. These rods were used to control the fission. ▼

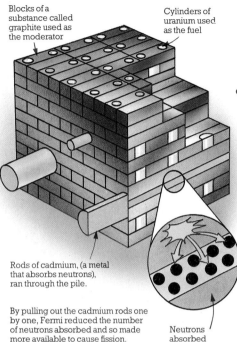

Blocks of a substance called graphite used as the moderator

Cylinders of uranium used as the fuel

Rods of cadmium, (a metal that absorbs neutrons), ran through the pile.

By pulling out the cadmium rods one by one, Fermi reduced the number of neutrons absorbed and so made more available to cause fission. Fission could be slowed down by pushing the rods back again.

Neutrons absorbed by cadmium rods.

- **1954** The first **reactor for generating electricity** was built in the USSR. It used graphite as the moderator and produced enough electricity for a town of about 6,000 people. The heat generated at the reactor core was used to boil water to make steam. This was used to drive **steam turbines*** in the electricity generator.

- **1956** The first **large-scale nuclear power station** was opened at Calder Hall, England. It was called a **magnox reactor**.

Nuclear fusion

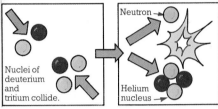

Neutron

Nuclei of deuterium and tritium collide.

Helium nucleus

The most powerful source of energy known is produced by **nuclear fusion**. When the nuclei of two **isotopes*** of hydrogen (deuterium and tritium) collide very fast, they fuse together to form a new element - helium.

- **1959** The first **fast breeder reactor** (FBR) was built at Dounreay, Scotland. This type is called a 'breeder' because it manufactures atoms at the same time as exploding them.

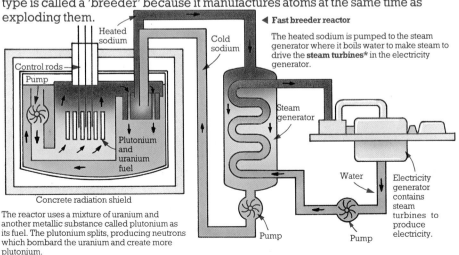

◀ **Fast breeder reactor**

Heated sodium

Cold sodium

Control rods

Pump

Plutonium and uranium fuel

Concrete radiation shield

The heated sodium is pumped to the steam generator where it boils water to make steam to drive the **steam turbines*** in the electricity generator.

Steam generator

Water

Electricity generator contains steam turbines to produce electricity.

Pump

Pump

The reactor uses a mixture of uranium and another metallic substance called plutonium as its fuel. The plutonium splits, producing neutrons which bombard the uranium and create more plutonium.

** **Electricity**, 6; **Isotope**, 118; **Nuclear bomb**, 92; **Steam turbine**, 9.*

Tools and machines

The axe, knife and hammer were the first tools to be made. The ancestor of human beings, homo erectus, was shaping stone axes in Africa, Asia and Europe by about 250,000BC. Tools were used by hand until the Industrial Revolution, which began in Britain in the 18th century. After this, many tools were operated by machines.

Woodworking and machine tools

- **c.3500BC** The earliest known **nails** were found in Iraq, in a statue of a bull made of copper sheets.

- **c.3000BC** The **bow lathe** was used in the Middle East to shape pieces of wood.

◀ **How a bow lathe works**
Bow string was wrapped round a spindle holding the object to be shaped. The bow was moved back and forth, turning the piece.
Sharp tool held against object to shape it.

- **c.3000BC** The **Egyptians** used **saws** to cut wood and stone. Saw marks can be seen on the stone of the pyramids.

- **AD79** The **Romans** used **planes** for shaving down wood. Examples have been found in the ruins of Pompeii.

A **Roman plane** had an iron ▶ blade set at an angle, protruding through a slit in the base.

- **c.1550** Metal **nuts** and **bolts**, for fastening things, appeared in Europe.

- **1775 J.Wilkinson** (GB) built the first accurate **water-powered machine**, for boring metal **steam engine*** cylinders.

- **1797 H.Maudslay** (GB) developed the ancestor of the modern **metal-working lathe**, an instrument for shaping metals. It was more accurate than earlier models.

Maudslay's metal-working lathe ▼

Modern lathes can shape extremely fine metal parts.

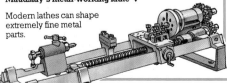

- **1839 J.Nasmyth** (GB) designed the first **steam hammer**, for beating out squares of metal. A heavy hammer head was lifted by steam pressure and then allowed to fall on the workpiece. This kind of hammer was used to produce the iron panels for Brunel's ship, the **Great Britain***.

- **1840s Pratt and Whitney Co.** (USA) introduced the **turret lathe**, which could combine a number of different cutting tools on the same lathe.

- **1950s Automatic machine tools** controlled by **computers*** were introduced in Europe and the USA. Computers can select and adjust the different tools.

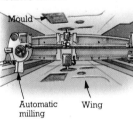

Mould ▶

Automatic milling machine   Wing

This picture shows an aeroplane wing being shaped by an **automatic milling machine**. The machine follows the pattern of the mould.

Potter's wheel

- **c.3500BC** The **potter's wheel** first appeared in Mesopotamia.

Early potter's wheel ▶

A flat table was spun round by hand. As it spun, the potter shaped the clay with his fingers.

Scissors

- **c.1000BC** The earliest **scissors** were in use in Europe and Asia.

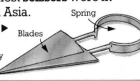

Spring

The **earliest scissors** were ▶ made of **bronze***. Overlapping blades were connected at the handle by a C-shaped spring.

Blades

* *Bronze*, 22; *Combine harvester*, 41; *Computer*, 49; *Diesel engine*, 9; *Great Britain*, 32;

Archimedean screw

- Between **287-212BC Archimedes of Syracuse** (Greece) designed the **Archimedean screw** for raising water from flooded ships. The principle is still used in irrigation in the Middle East and also in the grain unloader of the modern **combine harvester***.

◀ Archimedean screw
By turning a screw set in a wooden cylinder, the water is drawn up the screw and discharged at the top.

Pulleys

- **4th century BC** The first mention of a **pulley**, for raising heavy objects, occurs in a book from Greece called 'Mechanica'.

- Before **AD100 Hero of Alexandria** (Egypt) described a **compound pulley**.

When the rope is pulled down, the movable pulley and weight are raised through half the distance hauled on the rope.

Compound pulley ▼

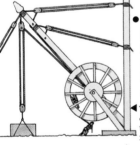

Fixed pulley

Movable pulley

Rope

Cranes

- **c.10BC Vitruvius** (Rome) wrote the first **description of a crane** in his handbook for architects.

◀ Vitruvius' crane consisted of a pole with a pulley at the top, held in position by ropes.

- **AD1805 J.Rennie** (GB) built the first **steam-powered crane**. Modern cranes are powered by **diesel engines*** and can reach to the top of **skyscrapers***.

Barbed wire

- **1873 J.Glidden** (USA) developed a machine for mass-producing **barbed wire** for cheap fencing.

Drills

- **1861 G.Sommeiller** (France) invented the first **pneumatic drill**, for use in building tunnels. Compressed air drove a piston which acted like a hammer, hitting a steel tool which smashed through the rock. Modern pneumatic drills work on the same principle.

- **1895 W.Fein** (Germany) invented the first hand-held **electric drill**.

Modern ▶ electric drill

Bearings and gears

- **c.1000BC** Wooden **ball bearings** were used in Europe to make the wheels of carts turn smoothly. They were located between the wheel hub and axle.

Modern ball bearings ▶

One race (ring) is fixed to the moving part of the machine, another is fixed to the stationary part. Steel balls between the two reduce friction. Stationary race

- **c.300BC Ctesibius of Alexandria** (Egypt) first used the **rack and pinion gear** in a **water clock***.

Rack and pinion steering* ▶

When a car steering wheel is turned, the rack moves from side to side making the front wheels turn.

To steering wheel

To front wheels

Pinion

Rack

Bunsen burner

- **1855 R.Bunsen** (Germany) designed the **bunsen burner**, a gas burner with adjustable air valve.

How a bunsen burner works ▼

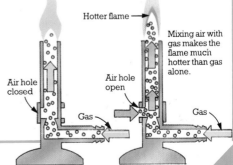

Hotter flame

Mixing air with gas makes the flame much hotter than gas alone.

Air hole closed

Air hole open

Gas

Gas

Rack and pinion steering, see **Steering box**, 31; **Skyscraper**, 24; **Steam engine**, 8; **Water clock**, 68.

Torch

- 1881 **E.Burr** and **W.Scott** (GB) patented the first **electric torch**. It was powered by a **wet-cell battery*** in a waterproof box.

Presses and jacks

- By **c.150BC** the **Greeks** had developed the **screw press** for making wine. By turning a handle, the end of a wooden beam was pressed down on a basket of grapes.

- **c.AD1250 V.de Honnecourt** (France) first illustrated the **screw-jack**. This is a hand-operated tool that can lift and support great weights with little human effort. The principle is still used to lift up cars in garages.

How a screw jack works ▼

By winding a handle, the jack is raised up while supporting huge weights. With a very fine screw thread, a small amount of effort can lift very large weights.

Handle

- 1795 **J.Bramah** (GB) patented the first **hydraulic press**, using the pressure of liquids. Modern hydraulic rams, used for raising and lowering the landing gear of aircraft, are similar in design.

Hydraulic press ▼

A slight pressure pushes the piston a long way into the smaller cylinder. The water then pushes up a much heavier weight in the large cylinder.

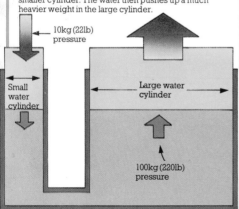

10kg (22lb) pressure

Small water cylinder

Large water cylinder

100kg (220lb) pressure

Elevators and escalators

- 1852 **E.Otis** (USA) invented a **hoist** to carry heavy machinery up floors in factories.

- 1857 **The Otis Steam Elevator Co.** (USA) installed the first **passenger elevator** in a shop in New York, USA. It could carry about six people. The invention of the elevator encouraged the building of **skyscrapers***.

Cutaway picture of an early Otis elevator ▼

- 1889 **The Otis Co.** (USA) installed the first **electric elevator**. It was lifted by cables driven by an **electric motor***.

- 1894 **J.Reno** (USA) designed the first **escalator**. It used the conveyor-belt principle to pull folding steps up a slope.

Gyroscope

- 1744 **The Royal Navy** (GB) made the first practical use of a **gyroscope**, a device that remains stable while spinning. It was used to indicate a stable horizontal reference for ships.

How a gyroscope works ▶

A flywheel is set spinning about an axis and remains on that axis. Everything connected to it also stays on the same axis.

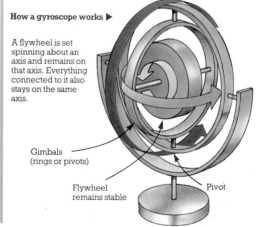

Gimbals (rings or pivots)

Flywheel remains stable

Pivot

Robots

- **1913 E.Sperry** (USA) introduced the **automatic pilot**, one of the first successful robots. A robot is an automatic device that can be programmed to do jobs. Instruments sensitive to the movements of **gyrocompasses** detect the changes in the height or direction of the aeroplane and activate the controls.

- **1940s Mechanically-operated arms** were developed in the USA to handle **atomic materials*** and dangerous chemicals behind a protective screen.

- **1962 Unimation Co.** (USA) sold the first **industrial robots**. They were simple machines that could pick up and move an object from one place to another.

- **1970s** More **flexible robots** were introduced in car factories in Europe, the USA and Japan. They could get behind and under car bodies for welding and spraying paint. One robot could do the work of many humans, at an even faster rate.

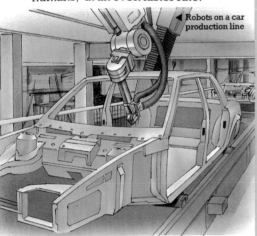

◄ Robots on a car production line

- By **1980 Unimation Co.** (USA) had introduced the **Puma robot**. It is sensitive enough for simple assembly tasks such as welding units and can position its 'fingers' very accurately.

Lasers

- **1958 C.Townes** and **A.Schawlow** (USA) suggested the **theory of laser light**, (Laser Amplification by Stimulated Emission of Radiation).

Laser beams are made up of light waves of identical wavelengths. A laser beam does not spread out like ordinary light waves and is therefore stronger.

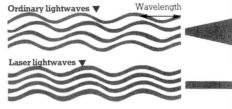

Ordinary lightwaves ▼ — Wavelength

Laser lightwaves ▼

- **1960 T.Maiman** (USA) built the first **laser**, a machine to make laser beams. It consisted of a rod of synthetic ruby crystal with a flash tube wrapped round it.

- **1960 D.Herriott, W.Bennett** and **A.Javan** (USA) developed the **gas laser**. It is not as powerful as the ruby laser but produces a continuous beam. (The ruby laser emits its beam in very short bursts). Lasers are used today in many different kinds of work, from very delicate surgical operations in **medicine*** to cutting metals.

How a ruby laser beam works ▼

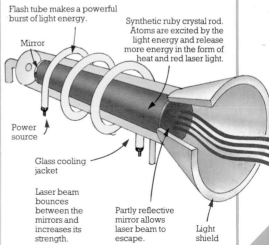

Flash tube makes a powerful burst of light energy.

Synthetic ruby crystal rod. Atoms are excited by the light energy and release more energy in the form of heat and red laser light.

Mirror

Power source

Glass cooling jacket

Laser beam bounces between the mirrors and increases its strength.

Partly reflective mirror allows laser beam to escape.

Light shield

---

* *Atomic materials, 16; **Electric motor**, 7; **Medicine**, 86; **Skyscraper**, 24; **Wet cell battery**, 6.*

21

Metals and synthetic materials

Some metals, such as gold, silver, copper and tin, occur naturally and are easy to work. They have been used to make objects for thousands of years. Others, such as iron, occur mixed with minerals in the form of ores which have to be heated to extract the metal. **Bronze** (copper and tin) was first made in about **3500BC** in Mesopotamia and started the first age of metals, called the Bronze Age.

Iron

●**c.1500BC** The **Hittites** of Anatolia (Turkey) first produced **wrought iron**. They burned iron ore (a mixture of minerals and iron dug from the ground), with wood and most of the impurities were then removed by repeated hammering.

●**c.600BC** The **Chinese** first produced **cast iron**. This was stronger than wrought iron. Better furnaces produced higher temperatures which could melt the iron completely.

●By **AD1400** the **blast furnace** was introduced in Holland. Blasts of air at the bottom produced very high temperatures, to make cast iron in Europe for the first time.

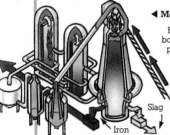

◀ **Modern blast furnace**

Hot air is blasted into the bottom half of the furnace, producing temperatures up to 1600° C.

Mixture of iron ore, coke and limestone are heated together. This produces pure iron and slag (impurities).

Slag

Iron

●**1947 H.Hartley** (GB) added a metal called **titanium** to iron to produce a much **stronger iron**.

Steel

●By **1000BC Steel** was being made in the Middle East and India. Steel is iron mixed with a small amount of **carbon** making it harder and stronger.

●**AD1856 Henry Bessemer** (GB) designed the **Bessemer converter**, a device that produced large amounts of steel cheaply.

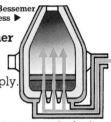

The Bessemer process ▶

Air is forced through holes in the base and through the molten iron, burning up the carbon.

A fixed amount of carbon is then added to make steel.

●**1864 W.** and **F.Siemens** (Germany) and **P.** and **E.Martin** (France) introduced the **Siemens-Martin open-hearth process**.

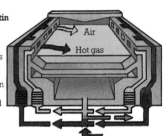

Siemens-Martin process ▶

Air

Hot gas

Air and hot gas pass over the molten metal. The gases from the molten metal are used to heat the air, so saving fuel.

●**1902 P.Héroult** (France) began producing steel in an **electric furnace**. This gave a very high temperature, producing much purer steel.

●**1913 H.Brearley** (GB) first made **stainless steel** by adding chromium to steel. This prevents rusting.

●**1948** The **basic oxygen process** was introduced in Austria. This is the main method of steel-making today.

Basic oxygen process ▶

A jet of oxygen is blown on to the molten iron, quickly burning up the carbon and producing steel. It is ten times faster than the open-hearth process.

Aluminium

- **1825 H. Oersted** (Denmark) first isolated **aluminium**. It comes from an ore called bauxite which contains the mineral alumina. He heated aluminium chloride (from alumina) with potassium.

- **1855 H.Deville** (France) improved the process by using **sodium**, which was cheaper than potassium.

- **1886 P.Héroult** (France) and **C.Hall** (USA) separately invented a process called **electrolysis** to divide molten alumina into aluminium and oxygen. This is the main process used today.

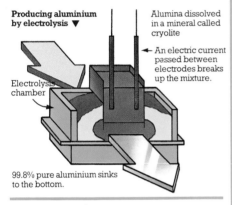

Producing aluminium by electrolysis ▼

Alumina dissolved in a mineral called cryolite

◄— An electric current passed between electrodes breaks up the mixture.

Electrolysis chamber

99.8% pure aluminium sinks to the bottom.

Synthetic materials

- **1862 A.Parkes** (GB) made the first **plastic** from cellulose nitrate, a substance made from natural plant cellulose mixed with camphor. It was a **thermoplastic**, one which melts when heated.

◄ **Thermoplastics** are made up of **molecules*** which are linked together in chains.

- **1869 J.Hyatt** (USA) worked on Parkes' method and produced a stronger plastic called **celluloid**.

- **1884 H.de Chardonnet** (France) produced the first **artificial fibre**. It was made from plant cellulose and was called **rayon** because of its shine.

- **1909 L.Baekeland** (Belgium) invented a plastic named **Bakelite**. It was made from Phenol (from coal tar) and a gas called formaldehyde moulded together under heat and pressure. It was the first material to be made from man-made chemicals and was a **thermosetting plastic**.

Thermosetting plastics are brittle and cannot be melted again once they have set. Their molecules are linked in all directions. ▶

- **1912 J.Brandenberger** (Switzerland) made **cellophane** from viscose (cellulose dissolved in sodium hydroxide and carbon disulphide). It is a thin plastic film used as a protective covering for food.

- **1927 I.G.Farbenindustrie Co.** (Germany) made the first **synthetic** (man-made) **rubber** from a **petrol*** chemical called butadiene.

- **1930 W.Chalmers** (USA) made **perspex** (plexiglass), a plastic 'glass' made from acrylic acid.

- **1935 W.Carothers** (USA) first patented **nylon**, made from a coal tar chemical called benzene.

Making nylon ▶

A solution of benzene was forced through tiny holes to form long fibres which were dried. These were then twisted into nylon yarn.

- **1941 J.Whinfield** and **J.Dickson** (GB) first produced **terylene**, used in textiles. It was made from two chemicals called terephthalic acid and ethylene glycol.

- **1943 PVC** (or **polyvinyl chloride**) was first commercially produced. It had been made earlier, but was too hard to use. It was made more pliable by adding chemicals called plasticers.

Building and building materials

The oldest known buildings in the world date from about 400,000BC and were discovered in southern France. They consist of traces of small huts made from branches.

Early building methods

- **c.10,000BC** The earliest stone buildings used a method of construction known as **post-and-lintel**. A stone lintel, or beam, lay across the top of two upright posts to form a doorway.

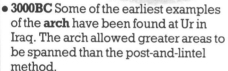

Stonehenge in ▶ England is the earliest remaining post-and-lintel building, dating from c.2000BC.

- **3000BC** Some of the earliest examples of the **arch** have been found at Ur in Iraq. The arch allowed greater areas to be spanned than the post-and-lintel method.

- **700BC** The earliest form of vault, known as a **tunnel vault**, has been found at Khorsabad in Iraq.

Tunnel vault ▶
An arch is lengthened into a tunnel which is supported by unbroken walls.

- **c.50BC** The **Romans** developed the **groin vault**, to let in more light.

A **groined vault** is ▶ formed when two tunnel vaults intersect at right-angles. Windows can be inserted in the sides.

- **AD120-124** One of the earliest examples of a **dome** was built in the **Pantheon**, Rome.

Inside the Pantheon ▶
The walls rise in a drum shape and carry a concrete dome covered in coffers, or sunken panels. Coffering helps to reduce the weight of the dome.

- **1162** The first use of the **flying buttress** was on **Notre Dame Cathedral** in Paris. It was an overhead strut that enabled higher walls to be built.

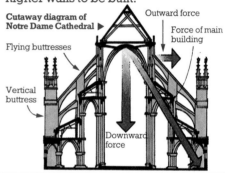

Cutaway diagram of Notre Dame Cathedral ▶

Outward force

Force of main building

Flying buttresses

Vertical buttress

Downward force

Modern building methods

- **1851 The Crystal Palace** in London was the first large **prefabricated building**. This is a building made up of modules or prefabricated (pre-made, off-site) units. These units can be reproduced many times to make up the complete building, which makes construction easier and quicker.

Crystal Palace ▼

3,300 identical **cast-iron*** columns were used.

- **1860** A **four-storey boat store** at Sheerness, England, was the first building with a completely **load-bearing iron* frame**. Very tall buildings need a frame to carry the weight of the building.

- **1884-85 Home Insurance Building** in Chicago, USA, was the first **skyscraper**. Its load-bearing frame was 52m (170ft) high and rose to ten storeys.

The **Sears Tower** in Chicago, USA, ▶ was completed in 1973 and is the tallest building in the world. It is 443m (1,454ft) high and rises to 110 storeys.

Bricks and tiles

- **c.6000BC** The oldest known **bricks** were used to build the city of Jericho in Jordan. They were made from clay and baked in the sun.

Sun-baked brick from Jordan ▼

- **3500BC Kiln-fired bricks** were first made in Mesopotamia. They were baked in kilns to about 1000°C. The extra heat made them very strong.

- **c.640BC** The earliest known **roof tiles** come from the Temple of Hera at Olympia in Greece and were made of fired clay.

- **AD1930s Breezeblocks** were first made in GB. They are made of furnace ash and cement and are cheaper but weaker than bricks.

Concrete

- **700BC** The earliest known example of **concrete** was used as the lining for an aqueduct at Jerwan in Iraq. Concrete is made of cement, sand and broken stones.

- **AD1867 J.Monier** (France) developed the first **reinforced concrete**.

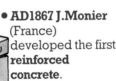

The concrete is very slightly flexible.

◄ In **reinforced concrete**, iron bars are embedded in the concrete to take the strain of the building.

- **1886 C.Dochring** (Germany) invented **pre-stressed concrete**. This is much stronger than other types of concrete.

How pre-stressed concrete is made ▼

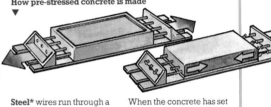

Steel* wires run through a mould and are fixed to two metal plates. The wires are stretched by pulling the plates apart. The concrete is then poured into the mould.

When the concrete has set the wires are cut off. Their high tension compresses the concrete slab and counteracts the weight of the building.

Glass

- **c.3000BC Glass** was first made in the Middle East by melting together sand (silica) and soda (sodium carbonate).

- **1st century BC** The **Syrians** introduced the **blowpipe** to make glass vessels.

- **c.AD1300 Crown glass** (flat glass) was in use in Europe.

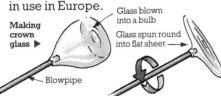

Making crown glass ►

Glass blown into a bulb

Glass spun round into flat sheet

Blowpipe

- **17th century** A new method of making flat glass, known as **plate glass**, was developed in France. Molten glass was poured on to an iron table to form a large sheet. Both sides were then polished.

- **1893 E.Libbey** (USA) introduced **fibreglass**, a material made from very thin strands of glass, which can be used to insulate buildings.

- **1952 A.Pilkington** (GB) developed the **float glass** process, which is the main method of producing glass today.

◄ Making float glass

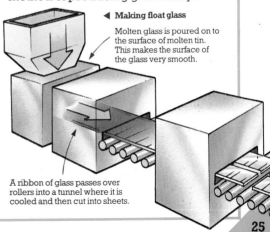

Molten glass is poured on to the surface of molten tin. This makes the surface of the glass very smooth.

A ribbon of glass passes over rollers into a tunnel where it is cooled and then cut into sheets.

*Iron, 22; Steel, 22.

Civil engineering

Roads

- **c.500BC** The **Persians** were the first to build a formal **system of roads** across their empire. They were made by pressing down the soil.

- By **c.350BC** the **Romans** had perfected their method of road-building. The skill of road-building was lost with the end of the Roman Empire.

Cross-section of a Roman town road ▼

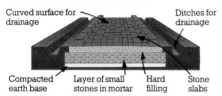

Curved surface for drainage

Ditches for drainage

Compacted earth base · Layer of small stones in mortar · Hard filling · Stone slabs

- **c.AD1750 P.Tresaguet** (France) designed the first **modern roads** using layers of stones, with cobbled surfaces.

- **Late 18th century J.McAdam** (GB) designed cheaper roads with very **smooth and waterproof surfaces**, formed from small stones.

- **1830s** A substance called **tarmac** began to be used on roads in Europe. It was made from **bitumen*** mixed with sand, which made the surfaces more flexible.

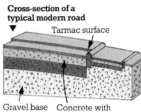

Cross-section of a typical modern road ▼

Tarmac surface

Gravel base · Concrete with steel mesh for reinforcement

Beam and cantilever bridges

- **c.10,000BC Beam** and **cantilever bridges** were in use throughout the world.

A **beam bridge** is made by laying a flat beam down on supports. The load is carried by the supports.

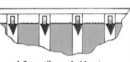

Central span

◀ A **cantilever bridge** is a refinement of the beam bridge. It enables the beam bridge to be used for longer spans. The central span pushes down through the supports and pulls up at each end.

- **AD1850 Robert Stephenson** (GB) completed the first **box girder bridge**, Britannia Bridge, carrying a railway over the Menai Straight in Wales. Here, the beams were rectangular tubes with trains running through.

Britannia Bridge ▼

Box girder bridges are often used today because they are light but very strong.

Suspension bridges

- **c.10,000BC** Primitive **suspension bridges** were used in the Stone Age. Suspension bridges can be made lighter and thinner than other designs.

The longest bridges are ▶ now all **suspension bridges** because they are much lighter. Most of the load is carried in the cables anchored to the banks.

- **AD1470** One of the **oldest surviving suspension bridges** is in Yunnan Province, China.

- **1801 J.Finlay** (USA) completed the first **modern suspension bridge** in Pennsylvania, USA. It had a stiffened deck to prevent it swinging around.

- **1966 G.Roberts** (GB) completed the Severn suspension bridge, England/Wales. The deck was made from **fin-shaped box sections**, which reduced wind resistance.

Severn ▲ suspension bridge

Streamlined box sections reduce wind resistance and allow a greater span.

Arch bridges

- **c.200BC** The **Romans** were the first to build **arch* bridges**.

  In an **arch bridge**, the load is ▶ carried at the ends of the arch.

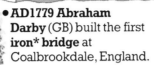

▼ Darby's iron bridge

- **AD1779 Abraham Darby** (GB) built the first **iron* bridge** at Coalbrookdale, England.

- **1874 J.Eads** (USA) completed the first **steel* bridge** at St.Louis, USA.

- **1901 R.Maillart** (Switzerland) designed one of the first **reinforced concrete* arch bridges** over the River Inn, Switzerland.

Canals

- **c.4000BC** The **earliest canals** were built in Mesopotamia to improve navigation by linking rivers together.

- **AD984 Chiao Wei-Yo** (China) designed the first proper **lock**.

  Locks on a canal are used to raise boats up or lower them from one level to another. ▼

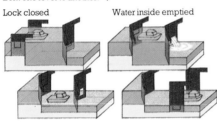

Lock closed

Water inside emptied

Water level lowered

Lock opened

- **1869 F.de Lesseps** (France) completed the **Suez Canal** in Egypt, designed to carry full-size ships.

  This is a ▶ **satellite*** picture showing how the **Suez Canal** links the Mediterranean to the Gulf of Suez.

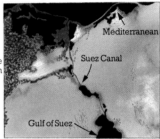

Mediterranean

Suez Canal

Gulf of Suez

- **1965** The world's **longest ship canal** was completed in the USSR. It is 2,977km (1,850 miles) long and links the Black Sea to the Baltic.

Tunnels

- **c.2160BC Queen Semiramis of Babylon** (Mesopotamia) built the earliest known **underwater tunnel**, under the River Euphrates.

- **AD1818 M.Brunel** (France/GB) designed the first **tunnelling machine**, used to build the Thames Tunnel in London.

◀ **Brunel's tunnelling machine** consisted of a wall of wooden planks. The planks could be removed one at a time to enable soil to be dug out. They were replaced and pushed further forward by jacks.

- **1970s Tele-Mole** (Japan) was introduced. It was the first **fully automatic tunnelling machine**.

Dams

- **c.3000BC** The **Egyptians** built the **oldest known dam** across the Garawi valley in Egypt. It was an **embankment dam**, built of earth and stones.

  The **Aswan Dam** in Egypt, ▶ completed in **1970**, is an example of a modern embankment dam.
  Water turbines* are used to generate **hydroelectric power***.

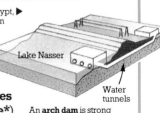

Lake Nasser

Water tunnels

- **c.AD560 Chryses** (**Constantinople***) completed the **Daras Dam** on the border of the Roman and Persian empires. It was an **arch dam**.

  An **arch dam** is strong because the load is transferred to the ends. ▼

* **Arch**, 24; **Bitumen**, 5; **Constantinople**, 118; **Hydroelectric power**, 10; **Iron**, 22;
Reinforced concrete, 25; **Satellite**, 38; **Steel**, 22; **Water turbine**, 10.

Trains and bicycles

The development of fast transport was made possible by the invention of the wheel. The earliest known picture of a vehicle wheel comes from Mesopotamia and dates from about 3200BC. It shows a cart with solid wooden wheels running on axles. Carts and carriages were the main form of transport on land until trains.

Trains

- **1803 R.Trevithick** (GB) built the first **steam locomotive** to pull wagons.

Trevithick's ▶ locomotive was driven by his **high-pressure steam engine***. This produced more power for its size than the larger factory engines.

- **1825 The Stockton and Darlington Railway** (GB) was opened. It was the first **passenger steam railway**.

- **1829 George** and **Robert Stephenson** (GB) designed a new engine called the **Rocket**, the ancestor of the modern locomotive.

Stephenson's Rocket had a new type of boiler, with 25 heating tubes leading to it from the firebox. These heated the water in the boiler to steam, which made the train powerful and fast. ▼

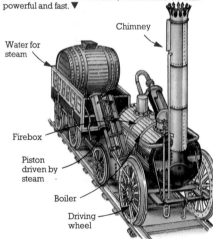

Chimney

Water for steam

Firebox

Piston driven by steam

Boiler

Driving wheel

- **1869 G.Westinghouse** (USA) patented the **air brake**. Compressed air forces pistons to clamp brake shoes against the wheels.

- **1879 W.von Siemens** (Germany) designed the first **electric train** using a live electric rail.

- **1881 W.von Siemens** (Germany) built the first **electric tramway** in Berlin, Germany.

A Siemens electric tram ▶ One of the rails carried the **electric current*** to the tram.

- **1912 Prussian-Hessian State Railways** (Germany) introduced the first **diesel*** **locomotive**. Diesel trains are cheaper to run than steam trains.

- **1971 Aérotrain** (France) was the first experimental **tracked air-cushion vehicle** (TACV) to be tested.

Aérotrain ran along a single rail on a cushion of air, ▲ as in a **hovercraft***.

- **1971 MBB** and **Krauss-Maffei** (Germany) demonstrated a prototype train with **magnetic levitation. Electric motors*** under the track created a **magnetic force*** which lifted the train 10mm (½in) above the track and pushed it along.

- **1981** Experimental **Advanced Passenger Train (APT)** (GB) was tested.

The **APT** tilts when ▶ cornering at speed to counterbalance the outward force. This means a constant speed can be maintained.

Underground trains

- **1863 Metropolitan District Railway** (GB) was opened in London. It was the first **underground railway** and used coke-burning steam locomotives.

Early London train ▶

- **1890 City and South London Railway** (GB) opened. It was the first **electric underground railway**. Travelling underground was now clean and comfortable.

Modern New York underground train ▶

Bicycles

- **1839 K.Macmillan** (GB) built the first **bicycle**.

Macmillan's ▶ **bicycle** was called a **velocipede**. The pedals were attached to rods which drove the back wheel.

- **1870 J.Starley** (GB) designed a **high-wheeled bicycle**, known as the **Penny Farthing**.

The Penny Farthing ▶

- **1885 J.K.Starley** (GB) brought together the main features of the **modern bicycle**. It had wheels of equal size with solid rubber tyres and gears.

Modern bicycle ▼

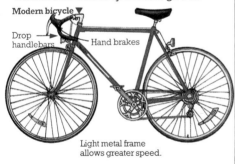

Drop handlebars
Hand brakes

Light metal frame allows greater speed.

- **1888 J.Dunlop** (GB) invented the **pneumatic cycle tyre** which was filled with air. This made cycling more comfortable.

Motorcycles

- **1868 E.** and **P.Michaux** (France) patented the first **motorcycle**.

The **Michaux** ▶ **motorcycle** was a bicycle with a small steam engine fixed behind the saddle. This was connected to the back wheel by leather belts.

- **1885 Gottlieb Daimler** (Germany) tested the first **petrol engine** on a wooden bicycle.

- **1895 Comte de Dion** and **G.Bouton** (France) designed a **lightweight engine** for motorcycles.

- **1901 M.** and **E.Werner** (France) made the first **modern motorcycle**.

◀ On the **Werner motorcycle**, the engine was fitted between the two wheels and controlled by grips on the handlebars.

▼ Modern motorcycle

In the 1950s motorcycles became more streamlined. Some of the latest motorcycles are capable of speeds of over 483kmh (300mph).

Diesel engine, 9; *Electric current*, 6; *Electric motor*, 7; *Hovercraft*, 33; *Magnetic force*, 7; *Steam engine*, 8.

Motor cars

The modern motor car, driven by an **internal combustion engine***, was not invented by any one person, but was the result of the work of many. An ancestor of the motor car was a steam-driven carriage built by N.Cugnot of France in 1770.

●**1885 Karl Benz** (Germany) produced the first **petrol-driven motor car**.

◀ **Benz's car** had three wheels and was powered by a belt-driven engine. Power was transmitted to the back wheels by a leather belt.

●**1887 Gottlieb Daimler** (Germany) built the first **four-wheeled car**.

Daimler's car looked like a carriage with an engine instead of a horse. ▶

●**1891 R.Panhard** and **E.Levassor** (France) produced the first car with the **layout of a modern car**.

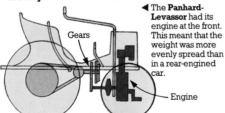

◀ The **Panhard-Levassor** had its engine at the front. This meant that the weight was more evenly spread than in a rear-engined car.

Gears

Engine

●**1895 E.** and **A.Michelin** (France) introduced the first **pneumatic car tyres**, which contained air. Earlier tyres were made of solid rubber.

●**1902 Renault** (France) introduced the modern **drum brake**. This had a better grip than the earlier band brake, which consisted of a steel band with curved shoes attached, tightened round the wheel hub by a lever.

Drum brake ▼
When the brake pedal was pressed, the hydraulic cylinder expanded, pushing the brake shoes against the inside drum of the wheel. This slowed the car down.

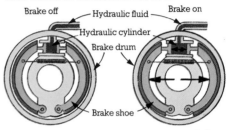

Brake off — Hydraulic fluid — Brake on
Hydraulic cylinder
Brake drum
Brake shoe

●**1902 F.Lanchester** (GB) introduced the more efficient **disc brake**.

Disc brake ▶
Brake pads squeezed against both sides of a steel disc inside the wheel, slowing the car down.

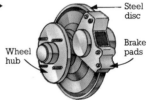

Steel disc
Wheel hub
Brake pads

●**1903 Spyker** (Holland) produced the first car in which all **four wheels braked**.

●**1908 Henry Ford** (USA) introduced the **Model T**, the first cheap, **mass-produced car**.

The **Ford Model T** ▶ was made of light but strong materials.

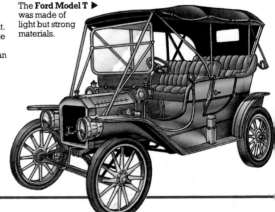

- **1914** The first **traffic lights** were installed in Cleveland, USA. They used red and green lights only.

- **1923 Studebaker** (USA) introduced cars with **all-steel* bodies**, a feature which became standard.

- **1934 P.Shaw** (GB) patented **cat's-eyes**.

◀ **Cat's eyes** are **glass prisms*** set down the middle of roads. They reflect car lights at night, so guiding the drivers.

- **1935 C.Magee** (USA) designed the first **parking meter**, installed in Oklahoma City, USA.

- **1940 General Motors** (USA) introduced **automatic transmission**, used in automatic cars. The gears change automatically without the driver having to select them.

- **1972 Dunlop** (GB) introduced **safety tyres**. A liquid sealing substance inside the tyre is released by the heat generated by a puncture and seals the hole. The car can be driven safely until the tyre can be changed.

How a safety tyre works ▼

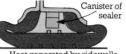

Deflation breaks lid off canister, releasing sealer.

Canister of sealer

Heat generated by sidewalls rubbing together.

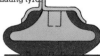

Heat vaporizes sealer, partly reflating tyre.

Layer of sealer inside tyre fills any small holes.

A modern car and parts

Discs apart - clutch disengaged.

Discs together - clutch engaged.

Clutch consists of two discs. The engine drives one disc all the time. The other one goes round only when it is pressed close to the first one. The clutch is then 'engaged'. When the clutch is disengaged, the engine is not connected to the wheels, so although the engine goes the car doesn't move.

Carburettor, where the petrol is mixed with air before passing into the cylinders.

Engine* makes the mainshaft turn round, which drives the back wheels.

Battery* provides electricity to start the engine and work the lights.

Petrol tank. Pump keeps petrol flowing from here to carburettor.

Exhaust pipe lets out burnt gases from engine.

Steering wheel

Back axle

Radiator cools engine.

Steering box converts direction of steering wheel to wheels.

Gearbox. As a car drives along, the speed of its wheels varies. The gearbox enables the driver to balance the speed of the engine with the speed needed at the wheels.

Accelerator pedal. When this is pressed down, more fuel enters the cylinders and the engine goes faster.

Clutch pedal

Brake pedal. When the pedal is pressed, the piston in the master cylinder transmits the increased pressure to the front and back brakes.

Mainshaft from engine turns back wheels.

Differential. The motion of the mainshaft must be turned through a right-angle so that it drives the back axle. Two cogs are used for this.

Shaft from engine bevel with teeth at 45°.

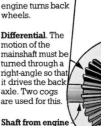

---

* **Battery**, 6; **Engine**, 8; **Glass prisms**, see **Binoculars**, 76; **Internal combustion engine**, 8; **Steel**, 22.

Ships

The earliest boats were made from natural materials such as wood, reeds and animal skins. The oldest known remains of a boat are those of a pine canoe found at Pese in Holland dating from about 7000BC.

Development of sailing ships

● **c.3000BC** The **Egyptians** used wooden ships with **single masts** for square sails.

▲ Egyptian sailing boat

Roman ship ▶

Extra sails

● **c.3rd century AD** The **Romans** improved on the Egyptian design by adding a **front sail** and a **topsail**.

● **9th century** The **Chinese** built ships with **several masts**. This increased sail area and speed.

Chinese ships also had submerged **rudders** which ▲ were stronger than oars and gave better steering.

● **15th century   Three-masted ships** were introduced into Europe. The shape of ships changed over the next 200 years.

◀ **c.1400** This typical boat, called a **Nef**, had 'castles' at both ends for soldiers and archers.

c.1550 A typical **galleon** had ▶ its front castle low down. Castles disappeared when cannon replaced archers.

● **1820s Clippers** were developed in the USA. They had **very large sails** and reached speeds of 20 knots (37kmh/23mph).

◀ **Clippers** were given their name because they clipped time off journeys.

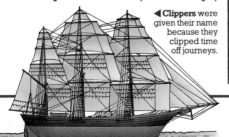

Iron and steam ships

● **1777** The first boat with an **iron hull** (body) was built in Yorkshire, England.

● **1783 Marquis de Jouffroy d'Abbans** (France) built the first working **steamboat**. It was made of wood and its **steam engines*** turned paddle wheels.

● **1836 F.Pettit Smith** (GB) patented the **screw propeller**, still used today. It was more efficient than the paddle wheel because it stayed under water and was less likely to be damaged.

● **1843 Isambard Kingdom Brunel** (GB) launched the **'Great Britain'**. It was the first **iron steamship with a screw propeller**.

The Great ▶ Britain could carry more cargo more safely than a wooden ship.

Modern ships

● **1886 Gottlieb Daimler** (Germany) introduced the **internal combustion engine*** in a boat.

● **1897 C.Parsons** (GB) introduced the **steam turbine*** engine, and fitted it to his boat, 'Turbinia'. This made it much faster than any other boat.

● **1902 'Petit-Pierre'** (France) was the first boat to carry a **diesel engine***. This is a more economic form of engine.

The Savannah ▼

● **1959 'Savannah'** (USA) was launched. It was the first merchant ship (for passengers and cargo) to be equipped with **nuclear power***.

Submarines

- **1620s   C.Drebbel** (Holland) built the first **submarine**, propelled by oarsmen. The body was made of greased leather over a wooden frame. The oars protruded through tight, greased-leather flaps.

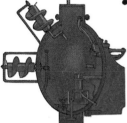

- **1776 D.Bushnell** (USA) designed a submarine called the **'Turtle'**, driven by a **hand-cranked propeller**.

◀ **The Turtle** contained buoyancy tanks which were flooded for submerging and pumped out for surfacing.

- **1863 'Plongeur'** (France) was the first **mechanically-driven submarine**. Its engine was driven by compressed air.

- **1900 J.Holland** (USA) designed the first **modern submarine**, 'Holland VI'. It was fitted with two engines: an **internal combustion engine***

for the surface and an **electric motor*** for underwater.

Holland VI could travel ▶ 800km (500 miles) on the surface and 40km (24 miles) submerged.

- **1940s** The **snorkel** was introduced in Germany. This is an air tube raised above the water which supplies air to diesel engines. Diesel engines could now be used while the boat was submerged.

- **1955 'Nautilus'** (USA) was the first **nuclear-powered* submarine**. In its first 2 years it travelled 99,800km (62,000 miles) without refuelling.

▼ Modern nuclear-powered Polaris submarine

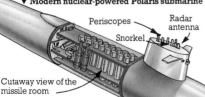

Periscopes   Radar antenna

Snorkel

Cutaway view of the missile room

Cutaway view of the torpedo room

Hydrofoils

- **1900 E.Forlanini** (Italy) built the first full-scale **hydrofoil**. A hydrofoil is a type of boat which is raised out of the water on foils (stilts). This reduces water resistance, enabling the boat to travel much faster.

Forlanini's boat was lifted ▶ on three sets of ladder foils and propelled by front and rear air propellers. At 64km/h (40 mph), only the bottom foil supported the boat.

The four main types of modern hydrofoil ▼

Ladder foils emerge from the water as the speed increases.

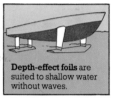

Depth-effect foils are suited to shallow water without waves.

Surface-piercing foils are the most common type for passenger hydrofoils.

Submerged foils are better for rough seas.

Hovercraft

- **1955 C.Cockerell** (GB) patented his invention of the **hovercraft**. The craft floats on a cushion of air, injected through a series of slits underneath.

Diagram showing how the modern **N4 hovercraft** works. ▼

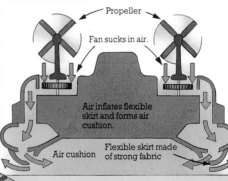

Propeller

Fan sucks in air.

Air inflates flexible skirt and forms air cushion.

Air cushion   Flexible skirt made of strong fabric

---

* *Diesel engine, 9; **Electric motor**, 7;*
***Internal combustion engine**, 8; **Nuclear power**, 16;*

Steam engine, 8; Steam turbine, 9.

Aircraft

Airships

- **1783 Joseph** and **Etienne Montgolfier** (France) launched the first passenger-carrying **hot air balloon**.

The **Montgolfier balloon** was ▶ made of paper-lined linen and was 15m (49ft) high. It travelled about 8km (5 miles) across Paris.

- **1852 H.Giffard** (France) designed and flew the first steam-powered **airship** for 28km (17 miles) at 8kmh (5mph). It was called a non-rigid airship because there was no frame inside the balloon.

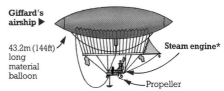

Giffard's airship ▶

43.2m (144ft) long material balloon

Steam engine*

Propeller

Aeroplanes

- **1804 G.Cayley** (GB) designed and flew a model of the first **true aeroplane**. It was a 1.5m (5ft) long glider. Cayley worked out the three basic principles of flight: **lift**, **thrust** and **control**.

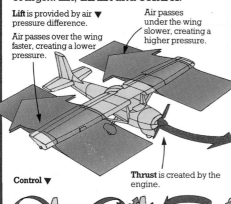

Lift is provided by air ▼ pressure difference.

Air passes over the wing faster, creating a lower pressure.

Air passes under the wing slower, creating a higher pressure.

Control ▼

Thrust is created by the engine.

The plane rolls by using its ailerons.

It pitches by using its elevators.

It turns by using its rudder.

- **1891-96 O.Lilienthal** (Germany) used a **hang glider** to make a series of controlled flights.

◀ **Lilienthal's glider** was controlled by leaning to and fro, as in a modern hang glider.

- **1903 Orville Wright** (USA) made the first **powered and controlled flights in an aeroplane**. It was called Flyer I.

Flyer I was a ▶ biplane (a plane with double wings) powered by an engine and two propellers. It covered a distance of 260m (852ft).

- **1909 Louis Blériot** (France) made the first **aeroplane crossing of the Channel**.

Blériot XI was a monoplane (a plane with a single wing). The 37km (23 mile) crossing took 36 minutes at an average speed of 64kmh (40mph). ▼

Engine in nose

Tail unit

Bracing to support wings

- **1912 Deperdussin racer** (France) was the first **monocoque aeroplane**. This meant it had a single body, rather than separate sections. It was made of wood.

- **1915 H.Junkers** (Germany) designed a series of monoplanes with thick, **self-supporting wings**, that needed no bracing.

Junkers JI ▶

- **1937 Lockheed XC-35** (USA) was the first **fully-pressurized aircraft**. This enabled greater speeds.

Jet aeroplanes

- **1939 H.von Ohain** (Germany) built the first **jet aeroplane** to fly. It was called a Heinkel He 178 and had two **jet engines***.

- **1947 Bell X-I** (USA) was the first aeroplane to exceed the **speed of sound*** (1,062kmh/660mph).

Bell X-I ▶

- **1949 De Havilland Comet** (GB) was the first **jet airliner** (passenger plane) to fly.

◀ The **Comet** had four **jet engines*** built into the wings.

- **1953 Rolls Royce 'Flying Bedstead'** (GB), an experimental aircraft, made the first **vertical take off** (VTO).

- **1961 Hawker-Siddeley P1127** (GB) was the first **VTO aeroplane**. It had a single engine which directed thrust down through four nozzles, two at each side.

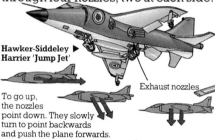

Hawker-Siddeley ▶ Harrier 'Jump Jet'

Exhaust nozzles

To go up, the nozzles point down. They slowly turn to point backwards and push the plane forwards.

- **1969 Boeing 747** (USA) first flew. It is the **largest civil airliner** in the world.

◀ The **Boeing 747** is designed to carry up to 490 passengers. This saves on flights and fuel.

- **1969 Concorde** (GB/France) first flew. It is the first **supersonic airliner**. Reaching more than 1,600kmh (1,000mph), it can cross the Atlantic in under three hours.

Concorde ▶

Hinged nose is lowered for landing

Helicopters

- **1907 P.Cornu** (France) succeeded in lifting a **rotating wing aircraft** 0.6m (2ft) off the ground. It was powered by an **internal combustion engine***.

- **1923 J.de la Cierva** (Spain) designed and flew a form of helicopter called an **autogyro**. He added a free-wheeling rotor (propeller) to a standard aeroplane with wings and propeller.

Cierva's autogyro ▶

Rotor

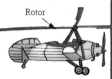

The rotor was not connected to the engine. The airflow caused by the craft's movement turned it.

- **1936 H.Focke** (Germany) designed the first **practical helicopter**, the Focke Fa-61. It had two counter-rotating blades set on arms.

- **1939 I.Sikorsky** (USSR/USA) designed the VS-300. This set the modern helicopter design of a **single main rotor** and **small tail rotor**.

Stabilizer fin ➔

Modern helicopter ▼

Main rotor

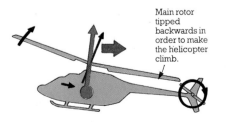

Tail rotor prevents the helicopter from spinning.

When in hovering flight, a combination of the speed of the rotor and the angle of the blades provides lift force, which balances the weight of the helicopter.

Main rotor tipped backwards in order to make the helicopter climb.

In order to climb, the rotor has to generate more lift. This is done by increasing the angle of the blades.

* *Internal combustion engine*, 8; *Jet engine*, 9; *Speed of sound*, see **Ernst Mach**, 116; *Steam engine*, 8.

35

Rockets and spacecraft

All rockets, from fireworks to spacecraft, work on the the principle of **Newton's third Law of Motion***: that every action has an equal and opposite reaction. When the fuel in a rocket burns, it produces gases which escape with great force in one direction and exert an equal force in the opposite direction, pushing the rocket up.

Development of rockets

- **c.1000** The **Chinese** made **fireworks** from **gunpowder***. (Gunpowder is a mixture of different chemicals that has to be ignited to explode.)

- **1232** The **Chinese** used **arrows attached to rockets** and fuelled by gunpowder to repel the Mongols at the Siege of Kai-fung-fu.

- **1804 W.Congreve** (GB) developed the first **rockets** to be used in European armies.

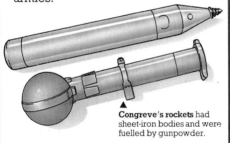

▲ **Congreve's rockets** had sheet-iron bodies and were fuelled by gunpowder.

- **1903 K.Tsiolkovsky** (USSR) published his **theory of rocket propulsion**, recommending the use of liquid fuels. When the fuels are mixed and ignited, they expand as a jet of gases.

- **1926 R.Goddard** (USA) launched the first **liquid fuel rocket**. It was about 3.05m (10ft) high.

Goddard's rocket had two fuel tanks, one containing petrol and the other liquid oxygen (which speeds up burning). A thrust was generated by the expanding gases.

▼ **Cutaway diagram of a rocket**

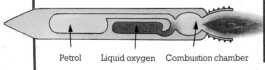

Petrol   Liquid oxygen   Combustion chamber

- **1942 Wernher von Braun** (Germany) launched the **V-2**, the first **long distance rocket**. It was the ancestor of nearly all space rockets.

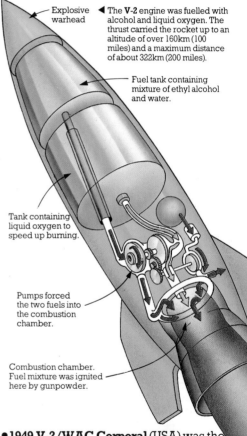

Explosive warhead

◀ The **V-2** engine was fuelled with alcohol and liquid oxygen. The thrust carried the rocket up to an altitude of over 160km (100 miles) and a maximum distance of about 322km (200 miles).

Fuel tank containing mixture of ethyl alcohol and water.

Tank containing liquid oxygen to speed up burning.

Pumps forced the two fuels into the combustion chamber.

Combustion chamber. Fuel mixture was ignited here by gunpowder.

- **1949 V-2/WAC Corporal** (USA) was the first **multi-stage rocket** to be launched. It consisted of two sections with one rocket each. This produced more power. The first stage and rocket was ejected soon after take off.

Going into orbit

The development of **multi-stage rockets** enabled people to travel out of the Earth's atmosphere and into orbit round the Earth. This requires a minimum speed of about 27,359kmh (17,000mph).

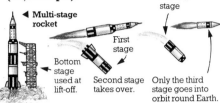

◄ **Multi-stage rocket**

Second stage

First stage

Bottom stage used at lift-off.

Second stage takes over.

Only the third stage goes into orbit round Earth.

●**1961 Yuri Gagarin** (USSR) became the first **person in space**. He completed an orbit of the Earth in Vostok I, in a flight lasting 108 minutes.

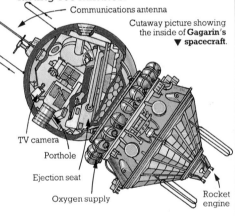

Communications antenna

Cutaway picture showing the inside of **Gagarin's** ▼ **spacecraft.**

TV camera

Porthole

Ejection seat

Oxygen supply

Rocket engine

●**1963 Valentina Tereshkova** (USSR) became the first **woman in space**. She made 48 circuits of the Earth in Vostok 6, in a flight lasting 70 hours 50 minutes.

●**1965 A.Leonev** (USSR) made the first **walk in space**. He spent ten minutes outside the two-manned Voskhod 2 spacecraft.

●**1966 N.Armstrong** and **D.Scott** (USA) made the first **docking in space** between a manned spacecraft and an unmanned space vehicle - Gemini 8 and Gemini Agena Target Vehicle.

Reaching the Moon

●**1964 Ranger 7** (USA), an unmanned **probe***, took **close-up pictures of the Moon**. It sent back 4,306 photographs, then crashed into the Moon.

●**1966 Luna 9** (USSR), an unmanned **probe***, **landed on the Moon**. It was the first mechanical object to land in working order and it sent back television pictures.

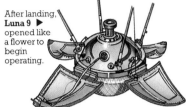

After landing, **Luna 9** ► opened like a flower to begin operating.

●**1968 F.Borman**, **J.Lovell** and **W.Anders** (USA) became the first **people to orbit the Moon**. They orbited 10 times in Apollo 8, coming within 112.6km (70 miles) of the surface, televising the pictures to Earth.

●**1969 N.Armstrong**, **E.Aldrin** and **M.Collins** (USA) travelled to the Moon in Apollo 11. Armstrong was the first **man on the Moon.**

Apollo Lunar spacecraft

Service module

Command module (the only part that returned to Earth)

Lunar module

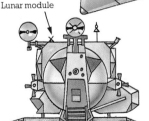

Collins stayed in the command module while Armstrong and Aldrin landed in the lunar module Eagle. They spent 2½ hours collecting rock and soil samples.

** **Gunpowder**, 90; **Newton's Third Law of Motion**, 81;*
***Probe**, 118.*

The conquest of space

Since space rockets were invented, satellites have been sent into orbit round the Earth, people have landed on the Moon and space **probes*** have been launched to explore distant planets. Experiments have been carried out in space stations.

Satellites

Satellites are unmanned spacecraft, launched by rockets, which orbit the Earth. There are about 300 in orbit today. They are used to send **telephone*** calls and **television*** pictures across the world and for **weather information*** and **navigation***.

- **1957 Sputnik I** (USSR), a satellite, became the first **vehicle to orbit the Earth**.

Sputnik I ▲

Exploring with space probes

- **1962 Mariner 2** (USA) was launched. It gathered **information about Venus**, finding a heavy atmosphere, mostly of carbon dioxide.

- **1964 Mariner 4** (USA) was launched on a mission to **Mars**.

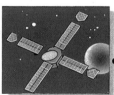

Mariner 4 came within ▶ 9,600km (6,000 miles) of the surface of Mars in 1965.

- **1970 Venera 7** (USSR) made the first **landing on Venus**. It sent back faint signals on atmosperic pressure and temperature.

- **1972 Pioneer 10** (USA) was launched on a 21-month mission to **Jupiter**. It was the **first man-made object to leave the Solar System***.

Pioneer 10 passed Jupiter in 1974. ▼

- **1973 Mariner 10** (USA) was launched on a mission to **Venus** and **Mercury**, taking the **first pictures of Mercury's surface**.

Deep craters on Mercury's ▲ surface

- **1975 Venera 9** and **Venera 10** (USSR) landed and sent back the **first pictures from the surface of Venus**. They gave readings on temperature and pressure.

- **1976 Viking I** (USA) **landed on Mars**. It sent back information on local conditions and looked for signs of life.

Viking I on Mars ▼

Communications antenna sent information back to Earth.

Cameras

Mechanical scoop carried soil samples back to craft.

Remote-controlled laboratory tested soil samples.

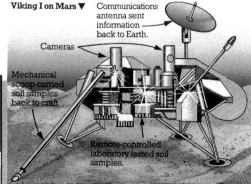

- **1977 Voyager I** (USA) was launched on a mission to **Saturn**.

- **1980 Voyager I** (USA) flew within 124,200km (77,000 miles) of **Saturn's cloud tops**.

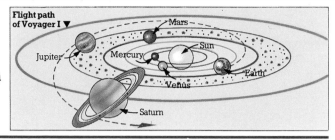

Flight path of Voyager I ▼

Mars · Jupiter · Mercury · Sun · Venus · Earth · Saturn

Space stations and experiments in space

- **1971 Soyuz 10** and **Salyut I** (USSR) docked in orbit to form the first **space station**. It was designed for research into **solar energy***, medicine and industrial processes.

- **1973 Skylab I** (USA) was launched for a range of **experiments in space**. It was visited by three crews during 1973 and 1974 and finally broke up in orbit in 1979.

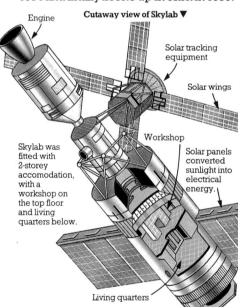

Cutaway view of Skylab ▼

Engine

Solar tracking equipment

Solar wings

Workshop

Solar panels converted sunlight into electrical energy.

Skylab was fitted with 2-storey accomodation, with a workshop on the top floor and living quarters below.

Living quarters

- **1975 Soyuz 19** (USSR) and **Apollo 18** (USA) docked in orbit for the **Apollo-Soyuz Test Project** (ASTP). It was the first joint USSR/USA space project and lasted two days, during which the crews carried out joint experiments.

Space shuttles

- **1977 Space Shuttle 'Enterprise'** (USA) made the first shuttle **test flight**. A space shuttle lands like an aeroplane and can be used many times, unlike a rocket. It is designed to carry scientific instruments into space.

Enterprise ▶ was carried up by a **Boeing 747*** mother ship.

- **1981 Space Shuttle 'Columbia'** (USA) was the first **space shuttle to go into orbit**. It orbited the Earth 36 times.

Stages of the Shuttle ▼

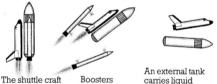

The shuttle craft and the two rocket boosters, are designed to be re-used, cutting the cost of space travel.

Boosters drop off at a height of 43km (26½ miles) and are recovered by ship.

An external tank carries liquid hydrogen and liquid oxygen for launch and climb. It drops off just before orbit and burns up in the atmosphere.

Space Shuttle ▼

Two orbital engines are used to move ship into, during and out of orbit.

Handling arm

Cargo bay open

Heat-resistant nose cap protects against 1,260° C entry.

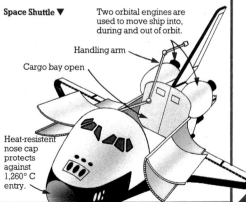

* **Boeing 747**, 35; **Navigation**, 72; **Probe**, 118; **Solar energy**, 12; **Solar system**, 118; **Telephone**, 52; **Television**, 58; **Weather**, 75.

Farming

The earliest evidence of a settled farming community is at Jericho in Jordan in about 8000 BC. Before that time, people had lived as nomads (travelling from place to place), hunting animals and collecting edible wild plants. Modern farming and food production is highly specialized, with scientific techniques.

Ploughs

- **c.3500BC** The earliest known picture of a **plough**, an instrument for turning over the soil, is shown on a seal from Ur in Iraq.

- **c.500BC** The **iron share** (plough blade) appeared in Europe. It was much stronger and heavier than the wooden share and cut deeper into the soil.

- **10th century AD** The **wheeled plough** was in use in Europe. It was easier to pull and control than earlier ploughs.

Wheeled plough ▶

Wheels gave better steering

Mouldboard to turn over the soil

- **c.1730 J.Foljambe** (GB) introduced the **Rotherham plough**. It had a lighter but stronger wooden frame.

- **1789 R.Ransome** (GB) produced the first **all-iron plough**. This was the first mass-produced farm implement.

- **1832 J.Heathcote** (GB) patented a **steam plough**, powered by a **steam engine***.

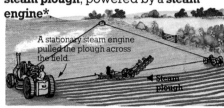

A stationary steam engine pulled the plough across the field.

◀ Steam plough

- **Modern ploughs** are pulled by tractors and have up to 12 bottoms. (A bottom is a ploughshare and mouldboard.) Each blade can rise independently.

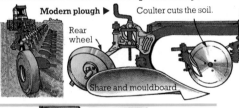

Modern plough ▶ Coulter cuts the soil.

Rear wheel

Share and mouldboard

Seed drills

- **c.1300BC** The **Babylonians** (Mesopotamia) had a **seed-dropper** for sowing seeds.

- **c.1570 T. Cavalini** (Italy) designed the first **seed drill**. It consisted of a hopper (container) mounted on a cart. The vibrations of the wheels shook the seeds through holes in the bottom.

- **c.1701 Jethro Tull** (GB) designed the first **efficient seed drill**.

The seed was dropped down tubes leading to the furrows in the soil.

Babylonian seed dropper ▶

Corn seeds were dropped down a tube.

Plough

Jethro Tull's seed drill ▼

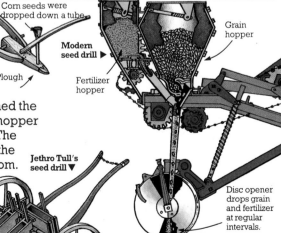

Modern seed drill ▶

Fertilizer hopper

Grain hopper

Disc opener drops grain and fertilizer at regular intervals.

** Gear speed, see **Gearbox**, 31; **Hydraulics**, see **Hydraulic press**, 20;*

Harvesting machines

- **1st century AD Pliny** (Rome) recorded that the Romans in Gaul (France) used a form of **reaping machine.** A cart with a comb of iron teeth at the front pulled the heads from stalks of wheat, which then fell into the cart.

- **1786 A.Meikle** (GB) designed the first practical **thresher**, a machine for separating the grain from the husk and stalks.

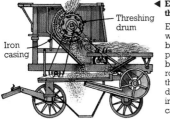

◀ **Early corn thresher**

Threshing drum

Iron casing

Ears of grain were beaten by being passed between a rotating threshing drum and an iron outer casing.

- **1826 Rev. P.Bell** (GB) designed the first practical **reaper** for cutting crops.

Bell's reaper ▶

Moving blades

Moving canvas band delivered cut crop to the side.

- **1836 H.Moore** and **J.Hascall** (USA) designed the first **combine harvester**, a combined reaping and threshing machine. It was horse-drawn and cut the crop and separated and bagged the grain.

Cutaway diagram of a modern combine harvester, showing the flow of grain and straw. ▼

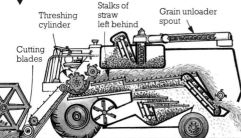

Threshing cylinder

Stalks of straw left behind

Grain unloader spout

Cutting blades

Milking machines

- **1862 L.Colvin** (USA) designed a hand-operated **milking machine**.

Early milking machine ▶

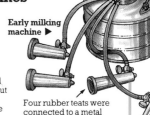

A **vacuum*** was produced by pumping and sucked out the milk. However, the continuous suction hurt the cow.

Four rubber teats were connected to a metal bucket.

- **1895 A.Shields** (GB) introduced the **Thistle milking machine**. It included a pulsating device which produced regular gaps in the **vacuum**, copying the sucking action of a calf. This feature is still used in modern milking machines.

Modern milking unit ▲

Milk from the milking unit is carried by pipe to churns or a milk tanker.

Tractors

- **1902 D.Albone** (GB) produced the first efficient **petrol-driven tractor.**

The **Albone tractor** had three **steel*** tyres. ▶

- **1916 Henry Ford** (USA) made the first **mass-produced tractors**.

- **1932 Rubber tyres*** were fitted to tractors. Before this, all tractors had steel tyres, with ridges to improve the grip.

- **1935 H.Ferguson** (GB) perfected a **hydraulic*** control system, so that ploughs and other implements could be controlled from the driving seat. Power is transmitted directly from the tractor's engine to the implements.

Modern tractors are ▶ very powerful and have up to 24 gear speeds*. (Cars have four to five.)

Steam engine, 8; *Steel*, 22; *Tyre*, 30; *Vacuum*, 118.

Fertilizers and food

In order to increase the amount of produce from their fields, farmers have always used some form of fertilizer and pest control. Until the invention of **refrigerators***, salting and smoking were the main methods of preserving food. Now there are ways of preserving food for months, such as drying and freezing.

Fertilizers

- **c.400BC Xenophon** (Greece) recommended a type of **green manure**, in which grass was ploughed back into the soil to keep in the goodness. Early fertilizers were also made from natural ingredients, such as animal droppings, blood and bones.
- **c.AD1840 J.von Liebig** (Germany) probably made the first **artificial fertilizer**. He treated bonemeal (ground animal bones) with sulphuric acid to produce concentrated **phosphate of lime**. Phosphate stimulates early root growth.

Healthy lettuce   **Lettuce with phosphate deficiency**, showing stunted growth and poor colour. ▼

- **1898 A.Frank** and **H.Caro** (Germany) made the first artificial **nitrogen fertilizer**. Nitrogen is an essential food for the growth of leaves and stems.

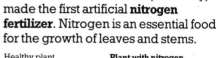

Healthy plant   **Plant with nitrogen deficiency**, showing small, pale leaves. ▼

- **1926 ICI Ltd.** (GB) began to make **compound synthetic fertilizers**, combining nitrogen, phosphate and potash (which regulates water).

- **1960s High nitrogen fertilizers** were introduced. They contain 82% nitrogen, injected as a liquid directly into the soil through sprays.

Liquid fertilizer spraying ▼

The soil and plants absorb the liquid fertilizers faster than powdered ones.

Pest control

- **c.2000BC** The **Chinese** used a form of **biological pest control**. Ants were placed on orange trees to kill leaf-eating insects.
- **AD1762 Biological control** was introduced in Mauritius when **mynah birds** were brought in to kill locusts.

Mynah bird with locust ▶

- **c.1763 Chemical control**, in the form of powdered **tobacco**, was first used in France to kill tiny insects called aphids.
- **1939 P.Müller** (Switzerland) discovered the insecticidal properties of **DDT**, a chlorine-based pesticide which was very effective against pests.

▼ **DDT being sprayed from the air**

DDT killed many other insects and birds and traces were found in humans. It is now banned in many countries. A variety of biological methods are being tested, such as sterilization of insects.

Bottled and canned foods

- c.1804 **N.Appert** (France) developed the technique of **bottling food.**

Bottling foods ▼

Jars filled with cooked food and corks put in loosely.

Jars are heated in boiling water, to remove any **bacteria*.**

corks knocked in tightly.

- c.1811 **B.Donkin** (GB) established the first **canning factory** for foods.

- 1855 **Mr.Yates** (GB) patented the first **can-opener**, a claw-shaped device for cutting open cans. Before this, cans had to be opened with a hammer and chisel.

Early can opener ▶

Dried and frozen foods

- **Early 19th century D.Edwards** (GB) produced **dried vegetables** by boiling them and then pressing them through small holes. The thin threads were then dried on heated plates. Drying food prevents it from rotting by keeping out the moisture that **bacteria*** need.

- 1901 **S.Kato** (Japan) produced the first **instant coffee**. Liquid coffee was boiled and evaporated and the residue was then dried to a powder. Instant coffee is now made by freeze-drying (see below).

- 1924 **C.Birdseye** (USA) set up a company to produce **frozen foods**. Freezing food slows down the process of decay. The faster the freezing, the better the food is preserved.

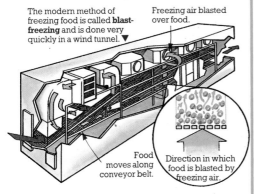

The modern method of freezing food is called **blast-freezing** and is done very quickly in a wind tunnel. ▼

Freezing air blasted over food.

Food moves along conveyor belt.

Direction in which food is blasted by freezing air.

- 1950s **Armour and Co.** (USA) dried food by **freeze-drying**. The food is frozen and then placed in a **vacuum*** chamber where the ice turns to gas.

Breakfast cereals

- 1829 **S. Graham** (USA) introduced the **first breakfast cereal**. It was a dry biscuit called the Graham Cracker.

- 1906 **W. Kellogg** (USA) first produced **Corn Flakes** commercially. This is the world's most popular cereal.

Artificial foods and flavouring

- 1869 **H. Mège-Mouriéz** (France) patented his recipe for **margarine**, a substitute for butter. It consisted of beef fat mixed with skimmed milk. Margarine is now usually made from vegetable oils such as soya oil.

- 1874 **W.Haarman** and **F.Tiemann** (Germany) produced the first **artificial food flavouring**. It was a vanilla essence made from the vanilla bean.

Today, many **artificial flavours** can be chemically re-created, from strawberries to smoky bacon. ▶

- 1930s **R. Boyer** (USA) worked out a recipe for **meat substitute**. Soya beans were ground to a flour, mixed with flavourings and spun into slabs. They could then be cut or minced like meat.

Soya plant ▼

Soya beans ▼

Meat substitute made from soya beans looks very like real meat. ▼

* **Bacteria**, 82 ; **Refrigerator**, 65; **Vacuum**, 118.

Writing

Sumerian pictographic inscription ▲

• **c.3500BC** The **Sumerians** (Mesopotamia) were the first people to write down their language properly. It is known as **pictographic** because it uses pictures to represent words.

• **c.3000BC** The **Persians**, **Assyrians**, **Babylonians** and other neighbours adapted this into a kind of writing, known as **cuneiform** (meaning 'wedge-shaped').

These pictures show the development of the **cuneiform** word for 'bird' from a pictograph. ▼

1.Pictograph for 'bird'   2.Turned to the left

3.Drawn as straight lines   4.Cuneiform for 'bird'

• **c.3000BC** The **Egyptians** introduced **hieroglyphs**. This is a system in which symbols represent words, syllables or sounds.

How hieroglyphs were used
▼ For example, this symbol could represent an eye, the word 'I' or the syllable 'i' as in 'island'.

• **c.1300BC** The first **true alphabet** evolved at Ugarit in Syria. Each letter represented a single sound which could be joined together to form a **complete word**. This alphabet is the ancestor of the European alphabet.

▼ **The Ugarit alphabet** contained 32 letters.

•**c.1300BC Chinese** writing evolved. Several hundred symbols were used, each representing a complete word. Chinese writing is similar today.

•**c.1000BC** The **Greeks** adopted the Ugarit alphabet. It developed into **two separate alphabets** - western and eastern Greek.

•**c.400BC** The **Etruscans** of Italy adopted the **western Greek alphabet**. This became the basis for the **Roman alphabet**, which is still used in the western world.

The development of the alphabet from Greek to Roman.

Greek

ΑΒΓΔΕ ΖΗΘΙ ΚΛΜΝΞΟΠ ΡΣΤΥ

Etruscan

A > ⱻꓞ Ɐ𐌀Ⱦ𐌏𐌉 𐌊ꓩꟿꟿⱭ 𐌕M𐌒4ᐅↆⱯ

Roman

ABCDEFG H IJKLMN OP QRSTUVW

Shorthand and braille

• **4th century BC Xenophon** (Greece) devised a **shorthand system**.

• **AD1784 V.Hany** (France) produced the first book with **raised letters**, designed to help blind people to read.

•**1824 L.Braille** (France) invented the **braille system** of embossed dots to represent words.

Braille alphabet ▼

A	B	C	D	E	F	G	H	I	J
K	L	M	N	O	P	Q	R	S	T
U	V	X	Y	Z	and	for	of	the	with
W									

•**1837 I.Pitman** (GB) published details of the **Pitman shorthand system**.

This means 'How do you do?'

▲ **Pitman shorthand** is based on the phonetic sound of words rather than their longhand spelling.

Writing implements and materials

- **c.4000BC** The **Egyptians** used hollow **reeds** as pens, with a type of ink made from soot mixed with water.
- **c.3500BC** The **Egyptians** used **papyrus**, a form of paper made from reed pulp.
- **c.1300BC** The **Greeks** used a **stylus** (pointed rod of bronze or bone) pressed into wax blocks to form letters.
- **c.500BC Quill pens** were introduced in Europe and the Middle East. They were made from sharpened feathers and were the main writing implements for more than 1,000 years.

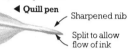

◀ **Quill pen**

Sharpened nib

Split to allow flow of ink

- **1st century AD** The **Chinese** made the first **true paper** from rags, wood and straw mixed with water and beaten together. The technique was used in Europe until the 19th century.

- **1565 K. Gesner** (Germany) first described a **pencil**. It was a piece of lead held in a stick of wood.
- **1799 N.Robert** (France) patented the first **papermaking machine**.
- **1884 L.Waterman** (USA) produced the first workable **fountain pen**.

Fountain pens ▶ have reservoirs of ink to allow continuous writing.

- **1938 L.Biro** (Hungary/Argentina) patented the first modern **ballpoint pen**.

Cutaway view of a ballpoint pen ▶

Ink flows on to a free moving steel ball. ──

- **1950s Cartridge pens** were introduced. They are fitted with cartridges that are thrown away when empty.
- **1962 Fibre-tip pens**, made entirely of **plastic***, were developed in Japan. Ink flows through a fibre pad when it touches a writing surface.

Typewriters

- **1808 P.Turri** (Italy) designed the first known **typewriter**, to help blind people write more easily.
- **1874 C.Sholes** and **C.Glidden** (USA) produced the first **modern typewriter**.
- **1902 Blickensdorfer Co.** (USA) sold the first successful **electric typewriter**.

Modern electric typewriters are fitted with **silicon chips*** which enable them to carry out many jobs automatically. For example, they can feed in paper, arrange paragraphs, underline and correct words automatically. Some have memories so that whole texts can be typed out automatically.

The text appears on a display screen for checking before print out.

Plastic 'daisy wheel' in a cassette speeds up typing. Typeface can be changed by fitting a new cassette.

Word processors

- **1964 IBM** (USA) produced the first **word processor**. It was a typewriter with a **computer*** memory that stored the text on **magnetic tape***. When the tape was played back, the typewriter automatically printed out the text again.
- **1978 Qyx** (USA) introduced **magnetic discs**. These work on the same principle as **magnetic tape***, but allow quicker access to the stored information.

Modern word ▶ **processor**

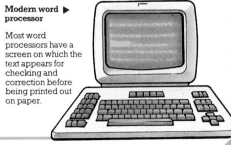

Most word processors have a screen on which the text appears for checking and correction before being printed out on paper.

Printing

Printing began in China. By 1300BC, the Chinese were using stone chops (seals engraved with a person's name or rank) for stamping documents. Letters were cut into the stone which was inked and pressed on to paper. Printed books developed from this.

● **AD868 The Diamond Sutra** (China) is the earliest known **printed book**. The areas to be reproduced were raised in relief above the areas to be left blank. This is known as **letterpress printing**.

A page ▶ **from The Diamond Sutra**

The book is in the form of a scroll and records the life and teachings of Buddha.

Letterpress printing ▼   Paper   Ink

Printing block. Design raised above areas to be left blank.

● **1040-50 Pi Sheng** (China) invented printing from **movable type**. Individual letters were arranged in an iron frame.

Printing presses

● **c.1450 Johannes Gutenberg** (Germany) invented the **printing press**.

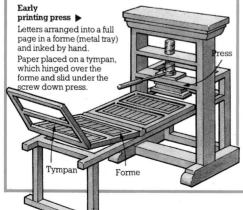

Early printing press ▶

Letters arranged into a full page in a forme (metal tray) and inked by hand.
Paper placed on a tympan, which hinged over the forme and slid under the screw down press.

Press

Tympan   Forme

● **1800 Lord Stanhope** (GB) invented the first **iron press**. It used levers instead of a screw mechanism, giving a firmer impression with less effort.

● **1812 F.Koenig** (Germany) made the first **cylinder press**. This has one flat and one cylindrical surface, with the printing plate on the flat surface.

● **1845 R.Hoe** (USA) designed the first successful **rotary press**. This design has two cylindrical surfaces. It prints from a roll of paper, with columns of type arranged round one cylinder.

Modern rotary presses are ▶ used to print newspapers. They can print up to 50,000 copies an hour.

Colour printing

● **1456 J.Fust** and **P.Schöffer** (Germany) invented **colour printing**. They inked different parts of the type separately in a second colour.

● **1719 J.Le Blon** (Germany) introduced the first full-colour printing process, known as **mezzotint**. He used red, yellow and blue to make each print.

Modern four-colour process ▶

The subject is photographed four times through four different coloured filters. Each filter separates out one basic colour. From these photographs four plates of the subject are made. The plates are printed on top of each other with different-coloured inks.

The final image combines the four colours.

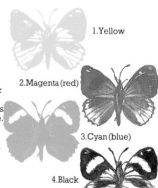

1.Yellow

2.Magenta (red)

3.Cyan (blue)

4.Black

Lithography

- **1798 A.Senefelder** (Czechoslovakia) invented **lithography**, a method of printing pictures from a stone surface. The image was drawn in reverse with a greasy crayon, then dipped in water. Oily inks and crayons repel water. Ink attached itself only to the parts which had been covered in crayon.

In **modern lithography**, a metal plate is used instead of stone and the image is formed photographically. It is offset (transferred) on to a rubber blanket and then printed on to paper. ▼

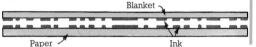

Blanket

Paper

Ink

Photogravure

- **1879 K.Klíc** (Czechoslovakia) introduced **photogravure**, a technique for reproducing illustrations. Small cells (holes) were etched with acid into a copper plate. Light areas in the image were etched with shallow cells, holding little ink; dark areas with deep cells, which held a lot of ink.

In **modern photogravure**, a photograph of the image is deposited on the plate. Cells are engraved by an automatic cutting head, controlled by a scanner that detects variations in the tone of the image. ▼

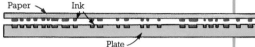

Paper

Ink

Plate

Typesetting

- **1886 O.Mergenthaler** (Germany) invented the **linotype machine**, the first machine able to set type automatically. It cast complete lines of type from molten metal.

- **1887 T.Lanston** (USA) invented the **monotype machine**, which is still used. It casts each letter individually.

◀ **Monotype typesetting**

The operator types the copy on a keyboard, producing a punched paper tape. The tape is fed into a second machine which casts individual letters, setting the letters in lines of equal length ready for printing.

- **1939 W.Huebner** (USA) invented the **photocomposing machine**, which sets type photographically on paper.

How a photocomposing machine works ▼

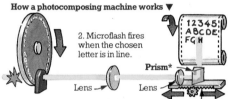

2. Microflash fires when the chosen letter is in line.

Prism*

Lens

Lens

1. Rotating disc carries negative images of all letters, numbers and punctuation marks.

2. Image focused on to photographic paper.

4. Transport system sets images in lines.

- **1965 Digiset**, a computer-controlled typesetting system, was invented in Germany. The text is stored in an electronic memory and then projected on to photographic paper. A **laser*** scans the text which is then transferred photographically to a printing plate.

Photocopying

- **1938 C.Carlson** (USA) patented his design for a **photocopier**.

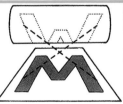

Photocopying ▶

An image of the document to be copied is projected on to an electrically charged plate. Light areas of the original destroy the **positive charge***. A positive charge is left where the original is black.

The plate is coated with **negatively-charged*** powder which sticks to the positively-charged (dark) parts of the plate.

The powder is transferred to a blank sheet of paper where it is sealed by heat.

Laser, 21; Negative charge, Positive charge, 14; Prism, see Binoculars, 76.

Mathematics and computers

- **c.3000BC** The **Sumerians** (Mesopotamia) devised one of the earliest **numbering systems**.

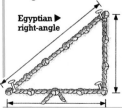

The Sumerian number system used **cuneiform***
symbols. The position of the symbols in the numbers indicated their value.

- **c.AD500** The **Indians** introduced **Hindu numerals**. Their symbols included a zero and were the ancestors of European numbers.

- **c.1200 Arabic numerals** were introduced into Europe by the Arabs, from India. They are the direct ancestors of European numbers.

Symbols for the numbers one to ten used by the Indians, Arabs and modern Europeans. ▼

Hindu	0	۱	۲	۳	۴	۵	۶	۸	۸	۹	
Arab	•	١	٢	٣	٤	٥	٦	٧	٨	٩	
European	0	1	2	3	4	5	6	7	8	9	10

Algebra, logarithms and binary arithmetic

- **c.1220 J.Nemorarius** (Germany) was one of the first people to use **letters to represent numbers**. This is the basis of **algebra**.

- **1614 J.Napier** (GB) published tables of **logarithms**. This enabled complicated calculations involving multiplication and division to be worked out by simpler addition and subtraction.

How logarithms work ▶

$$3 \cdot 89 \times 2 \cdot 54 = ?$$

The logarithm for 3.89 is 0.5899 and for 2.54 is 0.4048. To multiply the original numbers you add their logarithms, to make 0.9947. This is the logarithm for 9.879 which is very close to the exact answer of 9.8806.

- **c.1622 W.Oughtred** (GB) made the first **slide rule**, a device to simplify multiplication and division. It was a by-product of the invention of logarithms.

- **1679 G.Leibniz** (Germany) worked out the idea of **binary arithmetic**. This is a number system which uses a base of 2, instead of 10, in which any number can be represented by using only two symbols, 0 and 1.

In the **decimal system**, ▶ which uses a base of 10, 571 means 5 hundreds + 7 tens + 1 unit.

100	10	1
5	7	1

512	256	128	64	32	16	8	4	2	1
1	0	0	0	1	1	1	0	1	1

In the **binary system**, 571 is represented by 1000111011, meaning 512 + 32 + 16 + 8 + 2 + 1.

Geometry

- **2600BC** The **Egyptians** used a detailed system of **geometry** to build the pyramid of Snefu. Geometry is the part of mathematics that deals with the relationship between lines, points, angles and surfaces.

Egyptian ▶ right-angle

The builders measured out the base of the pyramid with a rope which had 12 equal sections knotted in it. It was stretched around three pegs to form sides of three, four and five sections. This made a right-angle.

- **c.550BC Pythagoras** (Greece) worked out the **Pythagoras Theorem**. This states that the square of the hypotenuse (the longest side of a right-angled triangle) equals the squares of the other two sides added together. This can help to work out an unknown length of one of the sides of a right-angled triangle.

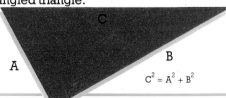

$$C^2 = A^2 + B^2$$

- **c.300BC Euclid** (Greece) outlined the **Greek knowledge of geometry** in his book, 'Elements', a standard school book for over 2,000 years.

- **c.AD1400 F.Brunelleschi** (Italy) discovered the geometric laws of **perspective**. This is the technique of representing distance on a flat surface.

- **1637 R.Descartes** (France) introduced **co-ordinate geometry**.

▲
In **co-ordinate geometry**, reference lines are drawn at right angles to each other. On a flat surface two lines are needed. Each point on a flat diagram can then be defined in terms of its distances from the two reference lines.

- **c.1666 Isaac Newton** (GB) established the rules for a **calculus**. He used this method of calculation to work out the curved path of a moving object in space, such as the **Moon's orbit round Earth***. The curve is known as differentiation and the area it encloses as integration.

Early calculating machines

- **c.3000BC** The **Chinese** or **Babylonians** (Mesopotamia) probably invented the **abacus**. It was one of the earliest and simplest forms of adding machine and is still used in parts of the USSR and the Far East. It consists of a wooden frame with rows of movable beads on rods. The beads represent units, tens, hundreds and thousands.

- **AD1642 Blaise Pascal** (France) built an **adding and subtracting machine**.

Pascal's machine was operated by a system of gears and wheels that could be turned to add up eight-figure ◄ numbers.

- **1674 G.Leibniz** (Germany) designed a machine that could also **multiply and divide**. But it was not reliable enough for general use.

- **1834 C.Babbage** (GB) designed an **analytical engine** which could store the results of calculations. But it was too complicated to build and was never completed.

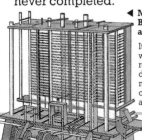

◄ **Model of Babbage's analytical machine**
It had rows of wheels representing the digits in a five-figure number and it was operated by turning a handle.

Computers

- **1945 P.Eckert** and **J.Maunchly** (USA) designed the **Electronic Numerical Integrator and Calculator (ENIAC)**. It was more like a giant calculator than a computer because it could not store data or programs.

- **1947 J.Bardeen**, **W.Brattain** and **W.Shockley** (USA) invented the **transistor**. This is a device that amplifies electronic signals like a **triode valve***, but is a fraction of the size and uses less power. It enabled much smaller computers to be made.

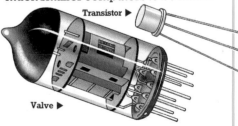

Transistor ▶

Valve ▶

- **1948 Manchester Mark I** (GB) was the world's first **proper computer**. A computer is a device that stores and processes information fed into it (input), producing the required results (output) very quickly.

- **Late 1940s Printed electronic circuits** were developed. The copper wires which carry the electric current are printed on to an insulating board.

Computers and silicon chips

- **1951 Ferranti Mark I** (GB) was the first computer to be **sold commercially**.

- **1958 Texas Instruments** (USA) produced the first **integrated circuit** or **silicon chip**. Each electrical component was now combined in one slice of silicon.

How silicon chips are made ▼

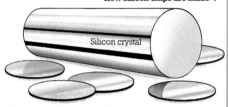

1. Crystals of silicon (a type of quartz) are grown artificially and cut into thin slices. Up to 500 chips can be made from each slice.

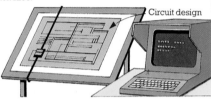

Circuit design

2. The circuit containing the electrical components for a chip is drawn out 250 times larger than it will be on the chip. The circuit design is then reduced to chip size and photographically copied up to 500 times on to each slice of silicon.

3. The slices are placed in an oven where they are exposed to different chemicals. Atoms of the chemicals enter the slices along the lines of the circuits. Each area of the chip is treated differently, according to its final function. These different parts are joined with fine wires to produce a complete electronic circuit.

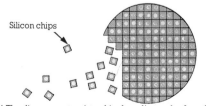

Silicon chips

4. The slices are cut up into chips by a diamond or **laser*** saw. Each tiny chip is then put in a plastic case.

- **1969 E. Hoff** (USA) invented the **microprocessor**. He placed all the circuits which do the work of the different parts of a computer on a single chip.

Enlarged picture of a microprocessor chip ▼

Chips like this are used in computers, calculators and in equipment such as **washing machines***, **typewriters*** and **sewing machines***.

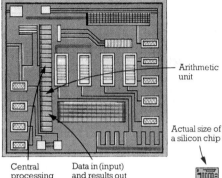

Arithmetic unit

Actual size of a silicon chip

Central processing unit

Data in (input) and results out (output) points

- **1971 Intel Corporation** (USA) introduced the first mass-produced **pocket calculators**, miniature computers that perform arithmetical calculations.

- **1975 Altair** (USA) produced the first **home computer**, a smaller, cheaper computer that could be used in private homes.

Modern home computer ▶

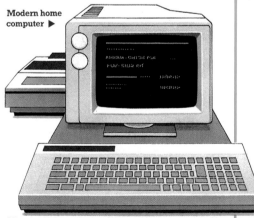

All sorts of information can be stored by computers and it can be called up quickly on the screen.

The telegraph and telephone

The telegraph, telephone, television and radio are all examples of telecommunications systems. They convert information into signals which can be transmitted over long distances by means of wires, radio waves or even glass fibres. The signals are then turned back into a copy of the original information.

Telegraph systems

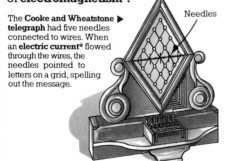

◀ Chappe's semaphore telegraph

● **1793-94 C.Chappe** (France) designed the first **semaphore telegraph**. Messages were sent from the top of a tower by means of waving metal arms.

The different positions of the arms stood for different letters.

● **1837 W.Cooke** and **C.Wheatstone** (GB) patented the first effective **electric telegraph**. It worked on the principle of **electromagnetism***.

The **Cooke and Wheatstone** ▶ **telegraph** had five needles connected to wires. When an **electric current*** flowed through the wires, the needles pointed to letters on a grid, spelling out the message.

Needles

● **1838 S.Morse** (USA) introduced **Morse code**. This uses signals transmitted electrically, consisting of short dots and long dashes to indicate letters.

● **1839** The first **commercial telegraph line** was installed in London, connecting Paddington and West Drayton.

● **1851** The first **underwater cable** was laid under the Channel between Dover, England and Cap Gris Nez, France. The first successful **transatlantic cable** was laid in **1858** between Ireland and Newfoundland.

● **1858 C.Wheatstone** (GB) patented an **automatic telegraph** system. Operators punched a message in Morse code on to a paper tape which then ran through a transmitter. A pen at the other end drew the signal on to another paper tape.

The **Wheatstone perforator** ▶ punched out telegraph signals on a paper tape.

● **1872 J.Stearns** (USA) patented **Duplex telegraphy**, which sent messages in both directions over the same line.

● **1931 American Telephone and Telegraph Co.** introduced a **teletypewriter exchange service**. A switchboard connected subscribers to the service.

● **1980s Teletex** system was introduced in Europe. In this, **wordprocessors*** thousands of miles apart can be connected. Letters written on a wordprocessor in one office can be immediately sent and printed out at their destination.

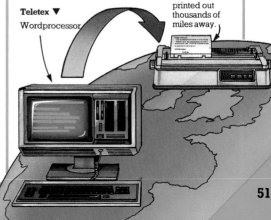

Teletex ▼
Wordprocessor

Same letter printed out thousands of miles away.

* Electromagnetism, 7; **Electric current**, 6; **Wordprocessor**, 45.

Telephones

•1876 Alexander Bell (GB/USA) patented the first **telephone**, which enabled speech to be transmitted along a wire. This telephone contained the first **microphone***.

Bell's telephone. A metal **diaphragm*** was placed next to an **electromagnet*** in the microphone. Talking into the microphone made the diaphragm vibrate. These vibrations were turned into a varying electric current by the electromagnet.▼

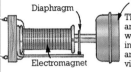

Diaphragm

Microphone

The current travelled down an iron wire to a receiver, where it was turned back into sound by making another electromagnet vibrate another diaphragm.

Electromagnet

•1878 The first **telephone exchange** was opened in Connecticut, USA. Telephone calls went via an operator who connected the callers to the people they wanted to talk to.

•1884 Bell Telephone Co. (USA) set up the first **long-distance telephone line** between Boston and New York. It used copper wire (instead of iron which rusted), which allowed the signals to travel further.

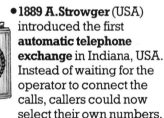

•1889 A.Strowger (USA) introduced the first **automatic telephone exchange** in Indiana, USA. Instead of waiting for the operator to connect the calls, callers could now select their own numbers.

Strowger ▲ telephone with dial (c.1900)

•1927 The first **transatlantic telephone links** between London and New York were opened. **Radio waves*** were used instead of wires to transmit the calls. One call at a time could be made.

• 1956 Tat 1, the first **transatlantic telephone cable**, was laid underwater between Scotland and Newfoundland. 36 telephone calls could be made at a time. The latest cable, the ninth, can carry 4,000 calls at a time.

•1960 Echo and **Courier satellites*** (USA) launched and relayed the first **satellite telephone calls** between the USA and Europe.

This is a picture of **Telstar** ▶ **satellite** (USA), launched in 1962. It was the first to relay live **television*** programmes as well as telephone calls.

•1966 K.Kao and **G.Hockham** (GB) first suggested using **fibre optic* cables** instead of copper wires to carry telephone conversations. Pulses of light are used to transmit the calls down the fibre optic cables. The first experimental **fibre optic telephone link** was opened in **1977** in Hertfordshire, England.

Fibre optic cable is made up ▶ of strands of glass. Each strand is the width of a human hair. Light is beamed into the inner core, bouncing along it. Thousands of telephone calls can be carried at the same time in each strand of glass.

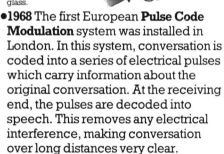

Light ray

Core

•1968 The first European **Pulse Code Modulation** system was installed in London. In this system, conversation is coded into a series of electrical pulses which carry information about the original conversation. At the receiving end, the pulses are decoded into speech. This removes any electrical interference, making conversation over long distances very clear.

•1979 British Post Office opened **Prestel**, a computer information service which can be connected to a **television set*** through a telephone.

Prestel ▶

Using a telephone, a range of information, such as banking services and travel, can be called up on a television screen from a central computer. Goods can even be ordered and paid for by dialling in a **credit card*** number.

*Credit card, 67; Diaphragm, 118; Fibre optics, 84; Microphone, 53; Radio waves, 53; Satellite, 38; Television, 58.

Radio and sound recording

●**1887/88 H.Hertz** (Germany) demonstrated the existence of a type of **electromagnetic wave**, called **radio waves**. A flow of **electricity*** in one circuit could result in a similar flow in another, separate circuit. This became the basis of the radio.

Hertz's experiment ▼

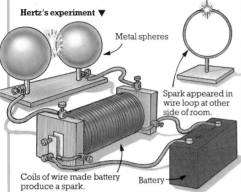

Metal spheres

Spark appeared in wire loop at other side of room.

Coils of wire made battery produce a spark.

Battery

●**1894 Guglielmo Marconi** (Italy) first tried to use **radio waves to communicate**, using Hertz's equipment.

Radio waves are pulses of electrical energy which travel through air, space and even solid objects.

Wavelength

A **wavelength** is the distance between one peak ▲ of the radio wave and the next.

The **frequency** of a radio wave is measured by the number of waves that pass a fixed point each second. The shorter the wavelength the higher the frequency.

Long wave radio has ▲ wavelengths of 1,000-2,000 metres.

Medium wave: 187-577 ▲ metres.

▲
Short wave: 10-100 metres.

Very high frequency ▲ **(VHF)**: 1-10 metres.

●**1901 Guglielmo Marconi** (Italy) sent out the first **radio signals across the Atlantic**, between Cornwall, England and Newfoundland, Canada. The signal was the **Morse code*** for S.

●**1906 R.Fessenden** (USA) made the first **radio transmission of speech**.

How sound is transmitted by radio
▼

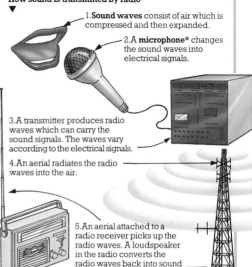

1.**Sound waves** consist of air which is compressed and then expanded.

2.A **microphone*** changes the sound waves into electrical signals.

3.A transmitter produces radio waves which can carry the sound signals. The waves vary according to the electrical signals.

4.An aerial radiates the radio waves into the air.

5.An aerial attached to a radio receiver picks up the radio waves. A loudspeaker in the radio converts the radio waves back into sound waves.

●**1906 L.de Forest** (USA) invented the **triode valve**, an electronic valve used to amplify electrical signals.

●**1924 E.Appleton** (GB) proved the existence of an area of electrified air 128-320km (80-200 miles) above the Earth called the **ionosphere**. It acts like a mirror, reflecting radio waves beamed into the air back down to Earth. This explained how radio signals travelled round the curved surface of the Earth.

Ionosphere ▼

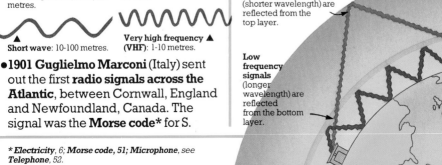

High frequency signals (shorter wavelength) are reflected from the top layer.

Low frequency signals (longer wavelength) are reflected from the bottom layer.

* Electricity, 6; Morse code, 51; Microphone, see Telephone, 52.

Radio cont.

- **1931 Microwaves**, radio waves under 30cm (12in) in length, were first used. They can easily be focused into sharp beams because of their short length. This reduces the chances of interference from other transmitters.

Sound recording

- **1859 C.Cros** (France) first **recorded sound waves** by engraving a layer of a soft material called graphite with a needle. However, he could not play back the recorded sound.

- **1877 Thomas Edison** (USA) designed a machine called a **phonograph** to record and play back sound. These were **monophonic recordings**, producing sound from one loudspeaker.

How Edison's phonograph worked

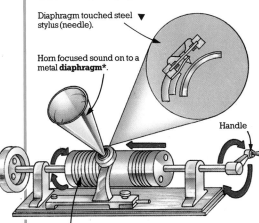

Diaphragm touched steel ▼
stylus (needle).

Horn focused sound on to a metal **diaphragm***.

Handle

Cylinder wrapped with tin foil

When the sound made the diaphragm vibrate, the stylus made indentations in the tin foil.

To play back, the cylinder was wound back to its original position. The indentations in the foil made the stylus vibrate and these vibrations were changed back into sound on the **diaphragm***.

- **1888 E.Berliner** (Germany) introduced the first **flat discs**, which replaced wax cylinders.

- **1954 Regency Co.** (USA) developed the first **transistor radio. Transistors*** replaced the large triode valves, allowing smaller radios to be made.

- **1961 Zenith** and **General Electric** companies (USA) developed **stereophonic sound** in radios. (See below for **stereophonic recording**).

- **1891 E.Berliner** introduced **master discs** from which copies could be made.

Producing a modern record ▼

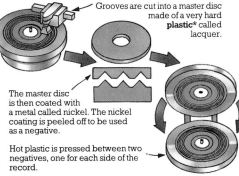

Grooves are cut into a master disc made of a very hard **plastic*** called lacquer.

The master disc is then coated with a metal called nickel. The nickel coating is peeled off to be used as a negative.

Hot plastic is pressed between two negatives, one for each side of the record.

- **1925 J.Maxfield** (USA) developed **electrical recording**. Instead of making a stylus vibrate, a **microphone** turned sound into electrical currents.

Cutaway diagram of a modern coil microphone. This type was first patented in 1877 in Germany. ▼

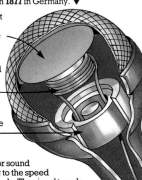

1. When sound waves hit this sensitive **diaphragm***, they make it vibrate at the same speed.

2. A coil of wire attached to the diaphragm vibrates at the same time.

3. Magnet. When the wire coil moves near the magnet, an electric current is produced.

4. The electric current, or sound signal, varies according to the speed and loudness of the sounds. The signal travels down a wire connected to the coil. It is then converted into mechanical force to operate a cutting tool which makes grooves on a blank master disc.

- **1933 EMI** (GB) developed **stereophonic recording**. This made the sound reproduction more realistic by recording with two microphones.

Stereophonic recording ▼

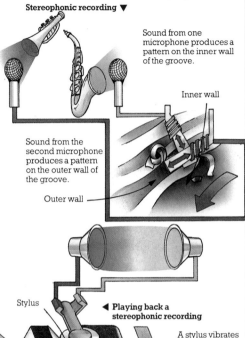

Sound from one microphone produces a pattern on the inner wall of the groove.

Inner wall

Sound from the second microphone produces a pattern on the outer wall of the groove.

Outer wall

Stylus

◀ **Playing back a stereophonic recording**

A stylus vibrates between the two walls of the groove. The sound played back through the two loudspeakers reproduces the original sound recorded into the microphones.

- **1971** Recording companies in GB and USA introduced **quadraphony**, with sound from four loudspeakers.

- **1982 Philips** (Holland) and **Sony** (Japan) produced the first audiodisc, called the **digital** or **compact disc**.

Compact disc   Standard LP

The compact disc is based on the same principle as the **videodisc***, in which the recording consists of tiny pits that are 'read' by a **laser beam***. A compact disc is only 12cm (4½in) in diameter and can play for an hour.

Tape recorders

- **1898 V.Poulsen** (Denmark) patented a machine for **magnetic* recording**. The varying pressure of the sound waves was converted into a magnetic pattern on a wire.

- **1928 F.Pfleumer** (Germany) introduced **magnetic tape**, made of paper coated with magnetic particles.

- **1935 AEG** (Germany) introduced the first **modern tape recorder**, using plastic tape coated with magnetic particles.

How a modern tape recorder works

Recording ▶

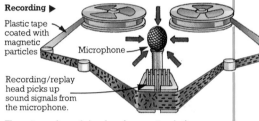

Plastic tape coated with magnetic particles

Microphone

Recording/replay head picks up sound signals from the microphone.

The pattern of sound signals makes a pattern in the position of the particles. Their magnetic strength varies with the strength of the sound signals.

Playing back ▼   Loudspeaker

The loudspeaker is driven by an electric current to make the sound.

Recording/replay head picks up the magnetic variations in the particles. It then reproduces the sound patterns that formed these variations.

- **1963 Philips** (Holland) introduced the **cassette recorder**, which uses small cassettes instead of large reels of tape.

Modern cassette recorders can ▶ be very small, taking miniature cassettes.

* *Diaphragm*, 118; *Laser beam*, 21; *Plastic*, 23; *Transistor*, 49; *Videodisc*, 59.

Photography, films and television

Development of cameras and film

- **c.1000** The **Arabs** discovered the principle of the **camera obscura**. This was a small, dark box with a tiny hole in one side that let in light from an object or scene and projected its image on to a screen inside the box.

Camera obscura ▼

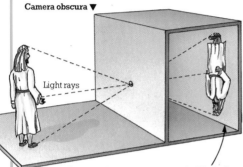

Light rays

Image projected on to a screen inside the box.

- **1558 G.della Porta** (Italy) replaced the pinhole with a **lens**, which increased the strength and brightness of the image.

- **1822 J.Niépce** (France) produced the first **fixed image**. A plate coated with a chemical called silver chloride reacted to light and left a permanent impression of the image it had been exposed to.

- **1830s L.Daguerre** (France) invented the first **practical photographic process**. The photographs, called daguerreotypes, were taken on copper plates coated with silver iodide. Only one print could be made from each shot.

Daguerreotype camera ▼

- **1835 W.Fox Talbot** (GB) invented the first **negative-positive process**. The photographs were called calotypes. Any number of prints could be made from the same shot by transferring the negative image on to special paper to make a positive print.

- **1851 F.Archer** (GB) invented the **wet collodion process**. He used a glass plate coated with a sticky substance called collodion. This cut the exposure time to a fraction of a second, enabling moving objects to be photographed.

- **1871 R.Maddox** (GB) invented the first fast exposure **dry plate**. The method used was basically the same as today's. Gelatine was used to bind together the processing chemicals.

Dry plate covered with gelatine containing grains of processing chemicals. The chemicals changed when exposed to light. ▼

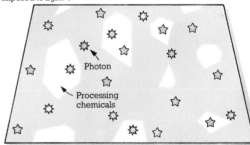

Photon

Processing chemicals

At least two photons (light particles) must hit a grain to form an impression.

- **1885 G.Eastman** (USA) introduced the first **roll film**. It was made of paper coated with light-sensitive paint.

- **1903 L.** and **A.Lumière** (France) produced the first **colour film**. Layers of red, green and blue grains were spread on to glass plates. These acted as filters, allowing some colours of light through and blocking others.

- **1925 Leitz Co.** (Germany) made one of the first small, **hand-held cameras**.

- From **1960s** many **modern cameras** are fitted with **silicon chips*** which enable them to carry out a variety of jobs automatically.

Modern automatic camera ▼

- **1947 E.Land** (USA) produced the **Polaroid camera**. It carries combined film and paper cartridges which incorporate the processing chemicals to develop the pictures. Today, some polaroid cameras use **sonar*** to focus their lenses.

Polaroid camera ▼

A sonar signal is sent out when the shutter button is released. The Signal hits the object and bounces back. A clock measures the time the signal takes to reach the subject and return. A **silicon chip*** in the camera uses this to calculate the distance of the subject and focus the camera.

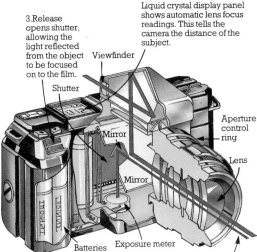

3.Release opens shutter, allowing the light reflected from the object to be focused on to the film.

Liquid crystal display panel shows automatic lens focus readings. This tells the camera the distance of the subject.

Viewfinder

Shutter

Mirror

Mirror

Aperture control ring

Lens

Batteries

Exposure meter

2.Some light is reflected downwards from a smaller mirror into the exposure meter. This calculates the amount of light available and sets the exposure - the amount of light allowed into the camera. This is controlled by the length of time the shutter stays open and size of aperture.

1.Light entering the lens is split into two paths. Most is reflected upwards from a mirror into a viewfinder. By looking through the viewfinder, you can see what the lens sees.

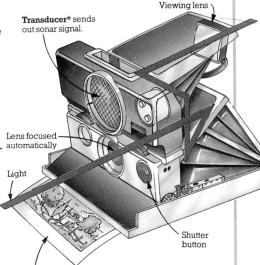

Transducer* sends out sonar signal.

Viewing lens

Lens focused automatically

Light

Shutter button

Photograph. Processing chemicals are spread on to the film as it emerges from the camera. It takes just a few minutes for the picture to develop.

Holography

- **1948 D.Gabor** (Hungary) invented **holography**, the technique of producing a three-dimensional photograph of an object by using **laser beams***.

Making a hologram ▶

The reference beam and reflected object beam meet at the plate. The two beams mix and make an 'interference pattern' in the light sensitive chemicals on the plate. This makes up the image of the object on the plate.

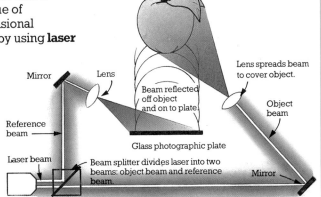

Mirror

Lens

Beam reflected off object and on to plate.

Lens spreads beam to cover object.

Object beam

Reference beam

Glass photographic plate

Laser beam

Beam splitter divides laser into two beams: object beam and reference beam.

Mirror

Laser beam, 21; Silicon chip, 50; Sonar, 73; Transducer, 118.

Motion pictures

- 1891 **Thomas Edison** (USA) patented his **kinetoscope**. It showed a series of photographs taken in sequence to give the impression of a moving picture.

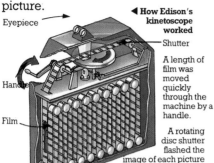

◀ **How Edison's kinetoscope worked**

Eyepiece

Handle

Film

Shutter

A length of film was moved quickly through the machine by a handle.

A rotating disc shutter flashed the image of each picture to the eyepiece.

- 1895 **L. and A. Lumière** (France) used a similar device in the first **public cinema**.

- 1923 **L.de Forest** and **T.Case** (USA) developed the **Phono Film system**. This was a method of matching the film sound to the picture, so that the sound was recorded directly on to the film.

- 1952 **CinemaScope** was introduced in the USA and Europe. This is a technique using a special lens to produce a very **wide picture**. The width is similar to the eye's field of vision and so gives a more realistic view.

How a movie projector works ▶

1. Feed reel feeds film into the projector. The scene is recorded as a series of pictures taken rapidly one after the other.
2. Gate. Each frame, or picture, is pulled down one by one into the gate and held for a fraction of a second.
3. A shutter behind the gate opens as each frame is stationary in the gate. It closes while the film is moved on to a new frame.
4. Light shines on to the film when the shutter is open.
5. The lens enlarges the picture on the film and focuses it on to the cinema screen.
6. A take-up reel collects the film at the end.

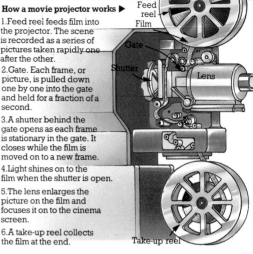

Feed reel
Film
Gate
Shutter
Lens
Take-up reel

How the soundtrack on movie film works ▶

A stripe along the edge carries the soundtrack. The width of this sound stripe varies according to the sound signals produced during the recording.

1. Light shines through the sound stripe. Because of the varying width of the stripe, a varying amount of light passes through to a device called a photoelectric cell.
2. A **photoelectric cell*** converts the light back into sound signals which are identical to the original sound signals.
3. The sound signals travel down a cable to the cinema's loudspeakers. These convert them into **sound waves***.

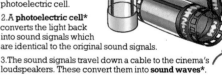

Sound stripe

Television

How Nipkow discs work ▼

- 1884 **P.Nipkow** (Germany) patented the **Nipkow disc**, used for transmitting an exact image of an object on to a screen.

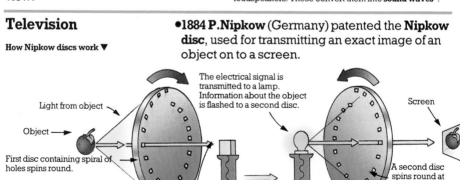

Light from object

Object

First disc containing spiral of holes spins round.

Light passes from the object through holes in the disc to the **photoelectric cell***. The cell converts the sequence of light into an electrical signal.

The electrical signal is transmitted to a lamp. Information about the object is flashed to a second disc.

Screen

A second disc spins round at exactly the same speed as the first. Light is projected through tiny holes in the disc on to a screen behind. The viewer's eye blurs all the points together and sees a complete image of the object.

- **1897 F.Braun** (Germany) invented the **cathode ray tube** which later became the basis for the modern television set. He painted the inside end of a glass tube with **fluorescent*** paint. A cathode (a type of **electrode***) inside the tube emitted **electrons*** which made the paint glow.

- **1925 V.Zworykin** (USSR/USA) patented the **iconoscope**. This was an electronic device with a lens which focused an image on to a screen inside a glass tube.

- **1925 J.Baird** (GB) gave the first **demonstration of television**. He used Nipkow discs to transmit images on to a screen, but the pictures were very blurred.

- **1936 British Broadcasting Corporation** began operating the first clear **black-and-white television service** from London. Baird's system was dropped in favour of one developed by Marconi and EMI (GB), based on the cathode ray tube and iconoscope.

- **1953** The first successful transmission of **colour television** was made in the USA.

How a picture reaches a television screen ▼

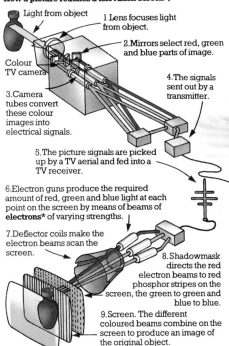

Light from object

1. Lens focuses light from object.

2. Mirrors select red, green and blue parts of image.

Colour TV camera

3. Camera tubes convert these colour images into electrical signals.

4. The signals sent out by a transmitter.

5. The picture signals are picked up by a TV aerial and fed into a TV receiver.

6. Electron guns produce the required amount of red, green and blue light at each point on the screen by means of beams of **electrons*** of varying strengths.

7. Deflector coils make the electron beams scan the screen.

8. Shadowmask directs the red electron beams to red phosphor stripes on the screen, the green to green and blue to blue.

9. Screen. The different coloured beams combine on the screen to produce an image of the original object.

Video tape recorders
- **1956 A.Poniatoff** (USA) introduced the first **video tape recorder**, called the Quadruplex. It recorded television pictures on **magnetic tape***.

- **1975 Sony** (Japan) introduced the first **domestic video tape system**, known as the Betamax. It uses a cassette to tape programmes while other stations are being watched.

Videodiscs
- **1970 Decca** (GB) and **AEG** (Germany) introduced a black-and-white **videodisc**, a television version of the normal audio record. A thin, flexible disc rotated on a cushion of air and was played with a needle.

- **1980s Philips** (Holland) introduced a colour **laser* videodisc system**. Lasers can produce clear recordings without wearing away the disc as a needle does. Videodiscs are similar to long playing records in size and shape and are cheaper to make than video tapes.

Videodisc ▼

To make a videodisc, pits are pressed into the disc. Their lengths carry information about the picture and sound.

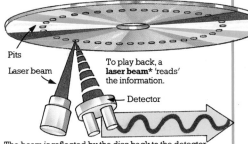

Pits

Laser beam

To play back, a **laser beam*** 'reads' the information.

Detector

The beam is reflected by the disc back to the detector which produces an electrical signal. This travels down a wire to a television set where it is converted back into a picture.

---

* Electrode, see **Battery**, 6; **Electrons**, 14; **Fluorescence**, see **Fluorescent light**, 62; **Laser beam**, 21; **Magnetic tape**, 55; **Photoelectric cell**, 110; **Sound waves**, 53.

59

Making cloth and clothes

The earliest clothes were made of animal skins and leather. With the invention of the spinning wheel, people were able to spin fibres such as wool and cotton into threads to make cloth.

Spinning machines

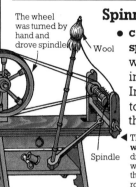

The wheel was turned by hand and drove spindle

Wool

- **c.AD1200** The **spinning wheel** was introduced into Europe from India. It was used to spin wool into threads.

◀ This is an **early spinning wheel**. Rotation of the wheel drew out fibres from a ball of wool and wound them into thread on a rotating wooden rod called a spindle.

Spindle

- **1764 J.Hargreaves** (GB) built the **'Spinning Jenny'**, a spinning frame with eight spindles which could be worked off one hand-turned wheel. More wool could be spun at once.

- **1769 R.Arkwright** (GB) patented the **water frame**, which made the process much faster.

Arkwright's loom was ▶ driven by a belt powered by a **water wheel***.

- **1779 S.Crompton** (GB) produced the **'Spinning Mule'**, combining the two previous inventions. One operator could spin up to 1,000 threads on 48 spindles.

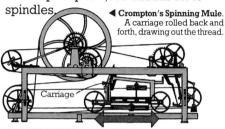

◀ **Crompton's Spinning Mule**. A carriage rolled back and forth, drawing out the thread.

Carriage

- **1792 E.Whitney** (USA) invented the **cotton gin**, the first machine to separate cotton fibre from the seeds. Before this, the work had had to be done by hand which was a very lengthy process. The principle is still used today.

Whitney's cotton gin ▶

Revolving hooks pulled the cotton through a grid but held back the seed pods.

- **1828 Ring-spinning** introduced in the USA, operating 3,000 spindles.

- **1965 Rotor-spinning** was introduced in Czechoslovakia. A high-speed rotor produces thread up to 6 times faster than ring-spinning machines.

Close-up picture of a ◀ **modern rotor-spinning machine**

Looms

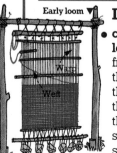

Early loom ▼

WARP

Weft

- **c.7000BC** The **earliest looms** consisted of simple frames holding one set of threads (the warp) while the weaver passed another thread (the weft) between them at right angles. The same design is still used in some places.

- **AD1733 J.Kay** (GB) invented the first **mechanized loom**, called the 'Flying Shuttle'. The shuttle holding the weft thread was shot very quickly through the warp threads by a cord controlling a shuttle box at either side.

- **1787 E.Cartwright** (GB) patented the first **steam-powered loom**, which made the process even faster. Modern looms are powered by **electricity***.

- **1801 J. Jacquard** (France) created the **Jacquard loom**, which could weave very detailed designs into the cloth. Perforated cards guided needles to lift warp threads, matching the pattern of holes on the cards.

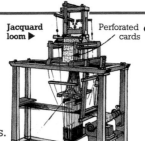
Jacquard loom ▶

Perforated cards

- **1950s Sulzer Co.** (Switzerland) introduced the **shuttleless loom**. Small projectiles are fired across the loom 320 times a minute. Each picks up the yarn and takes it through the warp.

Sewing machines

- **1830 B. Thimonnier** (France) designed the first true **sewing machine**.

Thimonnier's sewing ▶ machine

A needle rose and fell as a wheel was turned. A loop of thread was fed through the fabric by the needle and held by the next loop.

Wheel

Power from foot pedal (treadle)

Modern ▶ electric sewing machines can be fitted with **silicon chips*** which allow a wide range of different stitches and operations to be pre-programmed.

Fasteners

- **c. 2000BC Toggles**, buttons with string loops, were first used in the Middle East to hold clothes together.

- **AD1849 W. Hunt** (USA) invented the modern **safety pin** with a hidden point and coil spring.

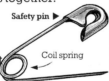
Safety pin ▶
Coil spring

- **1891 W. Judson** (USA) invented the first **zip fastener**. It interlocked by pulling a slide, but it was very unreliable.

- **1901 G. Abraham** (France) designed a **snap fastener**. Two small metal rings were pushed together, making fastening clothes quick and easy.

- **1906 G. Sundback** (Sweden) produced a **more efficient zip fastener** of metal teeth drawn together by a slide. This design is still used.

Sundback's zip fastener ▼

Clothes

- **c. 2500BC Shoes** (leather sandals) first appeared in the Middle East.

- **AD1200s** The **Amazonian Indians** made **waterproof material** by soaking it in sap from rubber trees.

- **1823 C. Macintosh** (GB) made the first **raincoats**, known as 'macintoshes'. He dissolved rubber in a substance called naptha and sandwiched it between layers of cloth.

- **1874 J. Davis** and **L. Strauss** (USA) used a strong cotton to make trousers known as **'jeans'**. They were called this because the cloth originated in Genoa, Italy.

- **1913 Caresse Crosby** (USA) made the first modern **brassiere**.

Crosby's brassiere ▶ consisted of two silk handerchiefs sewn together with ribbons.

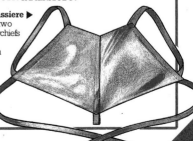

* *Electricity*, 6; *Silicon chip*, 50; *Water wheel*, 10.

Lighting and heating

Until about 150 years ago, the only form of lighting was by candles made of tallow (animal fat) or beeswax and by lamps burning animal or vegetable oils. The discovery of **petroleum*** led to the production of **paraffin*** as the main fuel for oil lamps. Later, **gas*** and then **electricity*** became the main fuels for lighting and heating.

Oil and gas lamps

- By **1784 A.Argand** (France) had designed an oil lamp with a **hollow wick.**

A current of air passed through the hollow wick, making the oil burn quicker and brighter.

▲ **Argand's oil lamp**

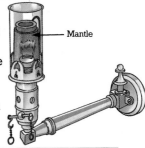

- **1885 C.Auer** (Austria) patented the first **gas mantle**. This made the light much brighter.

Mantle

Auer's mantle was a mesh of carbonized cotton that glowed brightly when heated in a gas flame. ▶

Electric lighting

- **1845 J.Starr** (USA) suggested using a **filament* of carbon** as an electricity conductor, enclosed in a glass bulb to produce a light.

- **1878 J.Swan** (GB) developed a **vacuum* pump** to remove air from the bulb so that the filament would not burn away.

- **1879 Thomas Edison** (USA) produced the first **electric light bulb.**

◀ **Edison's light** was a glass bulb containing a carbon filament in a vacuum. It burned for 13½ hours.

- **1908 W.Coolidge** (USA) used a **tungsten filament*** instead of carbon. It produced a much brighter light due to its high melting point.

Modern light bulb ▶

The coiled tungsten filament glows white hot when an electric current passes through it.

Vacuum prevents the filament from oxidizing and burning away.

Glass support column

Connecting wires to the electric current

- **1910 G.Claude** (France) produced the first **neon light**. An electric current was passed through neon gas in a glass tube, which made the gas glow red.

- **1935 General Electric Co.** (USA) first demonstrated **fluorescent lights**.

How a fluorescent light works ▶

Positive/Negative **electrode***

Tube filled with mercury gas

1. The filament of the negative electrode is heated and emits **negatively-charged* electrons*.**

2. The electrons are attracted to the **positively-charged*** filament.

3. An **alternating current*** is switched on, which makes the electrodes reverse roles continuously and causes the electrons to travel rapidly up and down the tube.

4. Some of the electrons collide with mercury atoms. This releases energy as ultra-violet light.

5. The inside of the tube is painted with a chemical called phosphor. The ultra-violet light makes the phosphor in the tube fluoresce (glow).

Negative/Positive **electrode***

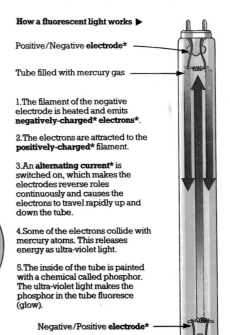

Central heating and air conditioning

Cold water storage tank

- **c.100BC** The **Romans** developed the first **central heating systems** for cold weather. But their ideas were forgotten with the end of the Roman Empire.

- **1716 M.Triewald** (Sweden) installed the first **hot water heating system** in a greenhouse in Newcastle, England. Water was heated by a fire and passed along pipes.

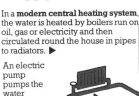

Hot water storage tank

In a **modern central heating system**, the water is heated by boilers run on oil, gas or electricity and then circulated round the house in pipes to radiators. ▶

Roman central heating ▼

Heat from a furnace was fed under the floors of houses through brick channels called **hypocausts** and through flues in the walls.

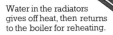

An electric pump pumps the water through the pipes.

Hot water

Cooler water

Pump

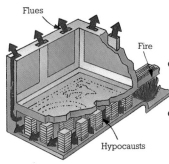

Flues

Fire

Water in the radiators gives off heat, then returns to the boiler for reheating.

Radiator

Boiler

Hypocausts

- **1902 W.Carrier** (USA) patented the first **air conditioner** in which air was blown across cold water. This controlled the temperature and humidity.

- **1906 S.Crawer** (USA) added **dust filters** to air conditioners. Today, air conditioners are essential in buildings such as **skyscrapers***, as they get very hot in the summer and very cold in the winter.

Gas and electric fires

- **1856 Pettit and Smith Co.** (GB) brought out the first **efficient gas fire** based on the **bunsen burner***. Air was drawn in below the flames, making them burn quicker and hotter than ordinary flames.

- **1892 R.Crompton** and **H.Dowsing** (GB) patented the first **electric fire**. An iron wire, protected in a coat of enamel, carried an electric current to an iron plate which then heated up.

- **1912 C.Belling** (GB) invented the first **efficient electric fire**. He used **nichrome wire**, a wire which could get very hot without burning up.

▲ Belling's electric fire

- **1937 Table fans** with **electric heaters** attached were sold in the USA. The fans blew air through the heaters and across the room.

- **1958 B.Eck** (Germany) designed the modern **fan heater**.

Cutaway diagram of a modern fan heater ▼

Cold air is blown over coils of hot wires to heat it, then blown out into the room.

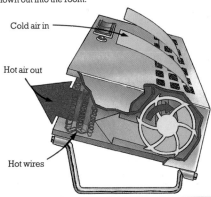

Cold air in

Hot air out

Hot wires

***Alternating current**, 118; **Bunsen burner**, 19; **Electricity**, 6; **Electrode**, see **Battery**, 6; **Electron**, 14; **Filament**, 118; **Gas**, 5; **Negative charge**, 14; **Paraffin**, 5; **Petroleum**, 5; **Positive charge**, 14; **Skyscraper**, 24; **Vacuum**, 118.

Domestic appliances

Housework today is much easier than it was 100 years ago. As more and more houses were fitted with electricity, hand-operated devices were replaced by ones driven by **electric motors***. Many appliances are now fitted with **silicon chips*** and can be pre-programmed to carry out different tasks.

Food mixers

- **1910 Hamilton Beach Co.** (USA) made the first **food mixer**.

◄ **Early food mixer** (1918)

- **1931 Flexible Shaft Co.** (USA) produced the first food mixer with a **built-in motor**.

- **1936 Sunbeam Corporation** (USA) developed the first **food processor**. It had attachments to peel, slice and liquidize fruit and vegetables, mince meat and grind coffee.

A **modern food processor** can be programmed to carry out a wide range of different jobs.▼

Hot drinks

- **1806 B.Thompson** (USA) invented the **coffee percolator**. Boiling water was poured over ground coffee in a metal filter and it dripped through.

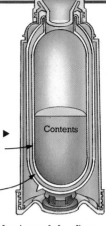

- **1892 J.Dewar** (GB) designed the first **vacuum* flask**, which could keep liquids either hot or cold.

How a vacuum flask works ►

Surface of inner glass wall is coated with mercury. This reflects heat or cold back to the contents.

Vacuum between the two walls prevents heat loss or gain by conduction.

Contents

- **1923 A.Large** (GB) designed the first **electric kettle**. It contained a copper element (wire) placed low in the kettle which carried the electric current.

- **1931 W.Bulpitt** (GB) invented a **safety plug** that sprang out and disconnected the power if the water overheated.

- **1955 Russell Hobbs Co.** (GB) designed the first **automatic kettle**. The hot steam cut off the power.

Modern automatic ◄ kettle

Toasting and frying

- **1909 General Electric Co.** (USA) produced the first **electric toaster**. It toasted bread, one side at a time, with red-hot wires.

- **1927 C.Strite** (USA) designed the first **pop-up toaster**. It had heating elements on both sides. A clockwork timer turned off the current.

Modern pop-up toasters ► can be pre-programmed to toast bread lighter or darker.

- **1958 M.Grégoire** (France) produced the first **non-stick frying pan**. He coated it with a **plastic*** unaffected by heat.

Cooking, freezing and washing

- 1802 Z.Winzler (Czechoslovakia) used the first **gas stove** for cooking food.

- 1879 K.von Linde (Germany) made the first **domestic refrigerator**. It was powered by a small steam motor which pumped round the cooling gases.

- 1889 The first **electric oven** was installed at the Hotel Bernina in Switzerland. The hotplates were slabs of solid metal heated by an electric current.

- 1899 Mrs. Cockran (USA) designed the first **dishwasher**. Dishes stacked in a tub were sprayed with water.

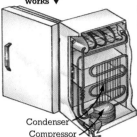

How a modern refrigerator works ▼

- 1923 B.von Platen and C.Munters (Sweden) designed the first **electric refrigerator**.

Freon gas is pumped by the compressor to the condenser. Here it condenses into a liquid and its temperature drops, cooling the refrigerator.

Condenser
Compressor

- 1953 Raytheon Manufacturing Co. (USA) patented the first **microwave oven**. High frequency **microwaves*** change direction very quickly, vibrating the water molecules inside the food and producing heat by friction.

- 1980s Thorn EMI (GB) developed **halogenheat**, in which food is cooked quickly by infra red light.

Infra red light waves are longer than ordinary **light waves*** and produce instant heat.

Halogenheat cooker ▲

Washing clothes

- 1782 H.Sidgier (GB) designed the first **washing machine**. It contained a cage of wooden rods turned by a handle.

- 1906 A.Fisher (USA) designed the first **electric washing machine**.

Control panel of a modern washing machine ▶

Ironing and sweeping

- 8th century The Chinese used some of the first **irons** to smooth silk. They were pans carrying hot charcoal.

- 1882 H.Seely (USA) invented the first **electric iron**.

- 1938 E.Schreyer (USA) invented the first **steam iron**. It had a thermostat (temperature gauge) which could be adjusted to different temperatures to suit different materials.

- 1876 M.Bissell (USA) patented a hand-operated **carpet sweeper** with a central rotating brush.

- 1901 H.Booth (GB) designed the first **electric vacuum cleaner**. An **electric motor*** drove a pump to suck up dirt.

Booth's vacuum cleaner ▼

Extinguishers and sprays

- 1816 G.Manby (GB) designed the first **fire extinguisher**. Compressed air forced out water from a metal cylinder.

- 1905 A.Laurent (USSR) designed the first **chemical fire extinguisher**, which smothered fires with a foam of aluminium sulphate and sodium bicarbonate.

- 1926 E.Rotheim (Norway) invented the **aerosol spray**. It used a pressurized gas to spray out liquids, such as insect repellent, as a mist.

How an aerosol spray works ▼

Mixture of product to be sprayed plus freon gas

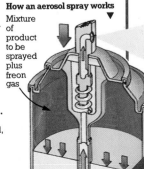

When the button is pressed, the pressure of the gas inside the can forces the mixture up the tube.

*Electric motor, 7; Light waves, see Laser, 21; Microwaves, 54; Plastic, 23; Silicon chip, 50; Vacuum, 118.

65

Locks and keys

- **c.2500BC** The **Chinese** probably invented the first **locks and keys**. They consisted of wooden bolts that were closed by pins fitting into slots.

- **AD1865 L.Yale** (USA) developed the **pin-tumbler lock**, one of the main types used today.

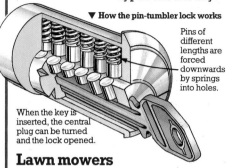

▼ How the pin-tumbler lock works

Pins of different lengths are forced downwards by springs into holes.

When the key is inserted, the central plug can be turned and the lock opened.

Lawn mowers

- **1805 T.Plucknett** (GB) designed the first **lawn mower**. It had a circular blade moved by two wheels. (Before this, people had cut grass with hand blades).

- **1902 Ransomes Co.** (GB) made the first **petrol-driven mower**.

Hover mower ▶

Modern lawnmowers include some that use the principle of the **hovercraft*** and even ones that can be operated by remote control.

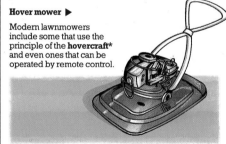

In the bathroom

- **c.200BC** The **Romans** built **public baths** and **lavatories** and used **taps**, but with the end of the Roman Empire their ideas were forgotten.

- **AD1589 J.Harington** (GB) designed the first **water closet**. It had a flushing cistern (tank for storing water) which drained into a pit below the house.

- **1775 A.Cumming** (GB) designed the first **modern lavatory**. It had most of the basic features used today.

Cumming's lavatory ▶

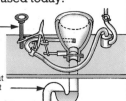

Pull-up handle opened valve releasing water from overhead tank. It also opened a side valve to let out contents of bowl.

U-bend with water. This cut off metal bowl from cesspit and stopped smells.

- **1780 W.Addis** (GB) designed the first **modern toothbrush**. It had a handle and bristles.

- **1800 T.Gryll** (GB) designed the modern **screw-down tap**. By turning a handle, you opened a valve which let in the water.

- **1840s The Duke of Devonshire** (GB) installed one of the first **baths** since Roman times at his house.

In the **duke's bath**, water ▶ from a tub was pumped up by hand into a shower head and cascaded onto the bather.

- **1868 B.Maughan** (GB) designed the first **water heater**, which used a gas burner.

- **1896 C.Braithwaite** and **E.O'Brian** (GB) designed a **valve** which fed gas to the burner only when the water tap was turned on. Modern water heaters are based on this.

- **1901 K.Gillette** and **W.Nickerson** (USA) patented the first **safety razor**.

- **1928 J.Schick** (USA) patented the first **electric razor**.

- **1960 Remington Corp.** (USA) produced the first **battery*-operated razor**.

Cutaway diagram of a ▶ modern battery razor

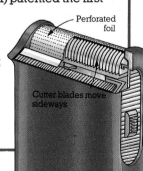

Perforated foil

Cutter blades move sideways

Money and postal services

Throughout history, many different things have been used for barter and exchange, such as copper bars, precious stones, shells and cattle. The invention of money freed people from the need to barter. The transaction could be made rapidly and simply by counting out coins.

Money

- **c. 700BC King Gyges of Lydia** (Turkey) probably issued the first **coins**. The idea spread quickly, through trade, around the Mediterranean.

Lydian coins were made ▶ from small lumps of a metal called electrum (a natural mixture of gold and silver). They were stamped on one side with the badge of the king.

- **6th century AD** The **Chinese** first issued **paper money**.

◀ Chinese banknote issued in about 1368.

- **1659 Messrs Clayton and Morris** (GB), bankers, handled the first known **cheque**. It was for the sum of £10.

- **1661** The first **European banknotes** were issued in Stockholm, Sweden. They were originally receipts issued by bankers for gold deposited with them, promising to repay the depositer.

- **1920s** The first **credit cards** were issued by USA oil companies for buying petrol.

- **1950 Diners Club Incorporated** (USA) introduced a **general credit card** for use in shops and hotels.

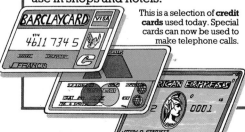

This is a selection of **credit cards** used today. Special cards can now be used to make telephone calls.

- **1980s Electronic cards** were introduced in France. This is the latest form of credit card and contains a **silicon chip*** which 'remembers' the holder's code number. A central **computer*** checks whether the card is valid and transfers payment.

Postal service

- **1464 Louis XI of France** set up a **state postal service,** the first since the Roman Empire, using messengers on horseback. The recipient of the letter had to pay for the postage.

- **1784** A **mail-coach service** was introduced in England, running between London and Bristol. Special coaches were built, pulled by horses.

Early mail-coach and ▶ horses

- **1830** Letters were first carried by **train** between Liverpool and Manchester, England.

- **1840** The **British Post Office** introduced the first pre-paid **postage stamp** for letters. It was called the Penny Black.

The Penny Black. For one penny, a 14.2 gm (½ oz) ▲ letter could be delivered anywhere in the country.

- **1852** The first **letter box** was put up in Jersey, Channel Islands. It was a tall, eight-sided iron box.

- **1911** The world's first **air mail service** began in Britain. Bags of letters were carried to the Continent by plane from London.

Computer, 49; Silicon chip, 50.

Clocks and watches

The earliest clocks

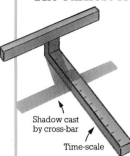

◀ Egyptian shadow clock

- **c.2000BC** The **Egyptians** were using **shadow clocks**.

The shadow of the cross-bar fell on a time-scale, indicating the different times. In the morning the clock was pointed east into the Sun, in the afternoon it was moved round to point west.

Shadow cast by cross-bar

Time-scale

- **c.1400BC** The **Egyptians** were using **water clocks**.

Egyptian water clock ▶

Water trickled through a hole in the bottom of a stone bucket. The time was indicated by the level of the water against a scale marked on the inside.

- **1st century AD** The **Romans** used **sand glasses**.

Sand glass ▶

When the sand in the top bulb had emptied into the bottom one, it meant that a fixed time had passed.

- **9th century King Alfred the Great of England** reputedly invented the **candle clock**.

◀ **Candle clock**

A candle was marked with hours. As it burned down, the time could be read off the scale.

- **16th century Oil clocks** were in use in Europe. The oil level indicated the time against a time-scale.

Mechanical clocks

- By **1300** fairly accurate **mechanical clocks** were made in Europe. They used a device called a **verge escapement**.

Inside a mechanical clock ▶

Balance moved back and forth and controlled the speed of ticking.

Crown wheel

The verge escapement consisted of a crown wheel and balance, connected through gears to the clock hands. As the weight fell, it set the crown wheel moving round in jerks or 'ticks'. This made the clock hands move round too.

Clock hands

Weight

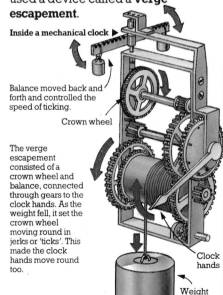

Pendulum clocks

- **1657 C.Huygens** (Holland) designed the first practical **pendulum clock**. The pendulum swung back and forth and regulated the ticking.

- **c.1660s W.Clement** (GB) replaced the verge escapement with an **anchor escapement**. This restricted the swing of the pendulum to a small arc, which improved the accuracy.

Inside a pendulum clock ▶

Each pendulum swing rocks the anchor. The anchor ends stop alternately the toothed wheel and each swing moves the clock forward one 'tick'.

Toothed wheel transmits 'tick' movement to escape wheel.

Escape wheel transmits 'tick' movement to clock hands.

Anchor

Pendulum

Weight

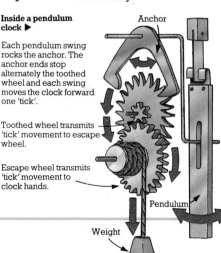

Spring-driven clocks

- **1675 C.Huygens** (Holland) designed the first clock with a **spiral balance spring**, making it very accurate.

- **c.1754 T.Mudge** (GB) introduced the **lever escapement**, used in most mechanical clocks and watches today.

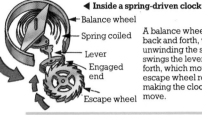

◀ **Inside a spring-driven clock**

- Balance wheel
- Spring coiled
- Lever
- Engaged end
- Escape wheel

A balance wheel moves back and forth, winding and unwinding the spring. This swings the lever back and forth, which moves the escape wheel round, making the clock hands move.

Watches

- **c.1500 P.Henlein** (Germany) made the earliest surviving **pocket watch**.

- **1790 Jaquet-Droz and Leschot Co.** (Switzerland) made the earliest known **wristwatch**.

- **1922 J.Harwood** (GB) invented the **self-winding wristwatch**. The mainspring was rewound by a weight that swung with the movement of the arm.

- **1956 Hamilton Watch Co.** (USA) produced the first **electric wristwatch** to be sold, powered by **batteries***.

Quartz clocks and watches

- **1929 W.Marrison** (Canada) produced the first **quartz crystal clock**. The mineral quartz has a piezoelectric effect. This means that if it receives electric charges it will vibrate at a fixed frequency and act like a pendulum. The vibrations are used to control the speed of an electric motor, which drives the clock hands. A quartz clock can keep extremely accurate time.

- **c.1967 Seiko Co.** (Japan) produced the first **quartz wristwatch**.

Diagram showing how a quartz clock works
▼

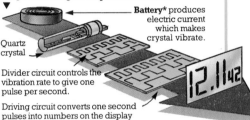

Battery* produces electric current which makes crystal vibrate.

Quartz crystal

Divider circuit controls the vibration rate to give one pulse per second.

Driving circuit converts one second pulses into numbers on the display panel.

Atomic clocks

- **1946 W.Libby** (USA) worked out the principle of the **atomic clock**, which uses vibrating **atoms*** of ammonia gas or of a substance called caesium. These atoms vibrate at a much faster rate than quartz crystal atoms. By using them, a quartz clock can be regulated with even greater accuracy.

- **1955 L.Essen** and **J.Parry** (GB) made the first **caesium atomic clock** to be used. It was accurate to one second in 300 years.

- **1969 Naval Research Laboratory** (USA) built the first **ammonia atomic clock**. It is accurate to one second in 1,700,000 years.

Diagram of an atomic clock ▶

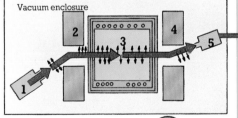

1. An oven heats the caesium in order to **magnetize*** the atoms.

2. Separator magnet. This allows only atoms with the correct magnetic alignment (position) to pass through.

3. **Microwave*** chamber. The atoms are mixed up again in here.

4. Selector magnet. Second selection process allows only atoms with correct magnetic alignment through.

Vacuum enclosure

5. An atom detector counts the atoms reaching it and produces an electrical signal based on that number.

6. Signal processing

7. Quartz clock. The electrical signal controls the vibration of the quartz crystal atoms to one pulse per second.

---

* **Atom**, 14; **Battery**, 6; **Magnetism**, 7; **Microwaves**, 54.

Weighing and measuring

In ancient times people used body measurements and many of the measuring units used today are based on them. The foot, for example, was originally the length of a Roman's foot. Nowadays very accurate instruments are used to determine standard lengths.

Earliest standards of length and weight

- **c.3000BC** The **Egyptians** used the **cubit**, a measurement of about 46cm (18in). It was the distance from the elbow to the tip of the middle finger.

Cubit measurement ▼

Stone cubit rulers were subdivided ▲ into smaller units.

- By **c.2600BC** the **Babylonians** (Mesopotamia) used the **mina**, the earliest known weight. It was a unit of about 907 grams (2lb).

Weighing instruments

- **c.4000BC** The **Sumerians** (Mesopotamia) invented one of the oldest surviving types of **balance**. It consisted of a beam with a central pivot suspended from a cord with ropes at each end.

- **c.1500BC** The **Egyptians** refined the balances used for weighing precious metals by bringing the scale pan cords out of the ends of a hollow beam. This meant that no matter how heavy the load, the cords were always against the ends of the beam. This produced more accurate readings.

◀ **Egyptian balance**

A suspension ring in the centre held a plumbline to indicate against a pointer when the beam was horizontal and balanced.

Scale pan cords

Pointer

Plumbline

Modern standards of length and weight

- **1588** The **yard** length and the **pound** weight were introduced in England. The word pound is derived from the Latin word meaning 'by weight'.

- **1793 The National Assembly** (France) introduced the **metric system**. It defined a metre as one ten-millionth of the distance from the North Pole to the Equator. One metre is now defined as 100cm.

- **1983 The International Conference on Weights and Measures** redefined the **metre** as the distance light travels during 1/299,792,458 of a second.

- **c.400BC** The **Etruscans** (Italy) probably invented the **steelyard**.

Pivot

◀ A **Roman steelyard**, based on the Etruscan design.

Scale pan

Weight

A weight was moved along the long arm until it was balanced. The distance the weight moved indicated the object's weight.

- **c.AD1500 Leonardo da Vinci** (Italy) described the first **self-indicating balance**. The weight of the load was shown by a fixed pointer attached to a beam read off against a scale.

- **1669 G.de Roberval** (France) invented the ancestor of the shop **counter scale**.

Roberval's counter scale ▼

The scale pans were mounted above the weighing beam, with no interference from chains and rods.

- **1718 J.Leupold** (Germany) built a **giant scale** for heavy loads such as hay wagons.

- **c.1770 R.Salter** (GB) invented the **small barrel spring balance**. It was inaccurate because the spring stretched, but very useful because it could be carried in the pocket.

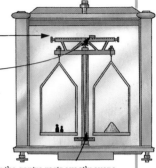

Salter's spring balance ▶ consisted of a strong spring inside a barrel marked with a scale.

Hook for attaching object to be weighed

- **1847 J.Béranger** (France) patented the **Béranger balance**, for weighing goods in shops.

Diagram of a Béranger balance ▼

Each scale pan was supported by a system of levers.

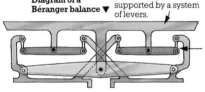

Shorter beams below the main double beam reduced friction, so improving accuracy.

- **1870s F.Sartorius** (Germany) patented a **very accurate balance**, which was accurate to one part in a million.

- **1980s** The most accurate **modern weighing balances** are enclosed in glass cases and controlled from the outside. This removes temperature changes and draughts which can disturb the balance.

Modern accurate balance ▶

Screw-weights for fine adjustment

Weight, designed to allow much finer measurements, slides along a scale on the top of the balance arm.

A long, vertical needle in the centre rests exactly over a central point on a scale at the base of the balance column. This shows the correct adjustment.

Measuring instruments

- **1638 W.Gascoigne** (GB) invented the **micrometer**, an instrument for measuring in astonomy.

- **1772 James Watt** (GB) probably made the first **screw micrometer**, an instrument for measuring the width of objects.

- **1848 J.Palmer** (France) made the first **modern screw micrometer** similar to ones used today. It is used for very accurate measurements in the engineering industry.

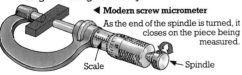

◀ Modern screw micrometer
As the end of the spindle is turned, it closes on the piece being measured.

Scale — Spindle

- **1973 James Neill Group** (GB) introduced the first **electronic micrometer**. It gives very accurate readings instantly on a digital display.

- **1976 NASA** (USA) launched the **laser geodynamics satellite* (LAGEOS)** for measuring very great distances accurately.

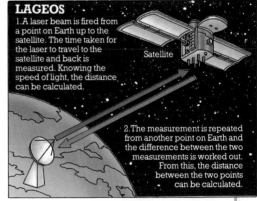

LAGEOS
1. A laser beam is fired from a point on Earth up to the satellite. The time taken for the laser to travel to the satellite and back is measured. Knowing the speed of light, the distance can be calculated.

Satellite

2. The measurement is repeated from another point on Earth and the difference between the two measurements is worked out. From this, the distance between the two points can be calculated.

- **1980s Sonar* measure** introduced. It is a hand-held device that sends out sonar beams from a **transducer***. The time the beams take to return from the object indicates the distance.

Surveying and navigation

Maps

- **c.2300BC** The **Babylonians** (Mesopotamia) produced the oldest known **map of the world**. It was a clay tablet showing the Earth as a small disc with Babylon at the centre.

- **c.AD150 Claudius Ptolemaeus** (Greece) produced an important **map and guide** of the known world. Until the 16th century, most maps were based on this.

- **1569 G.Kremer** (Holland), known as **Mercator**, produced the first world map which solved the problem of how to plot a **straight line on a curved surface**.

◀ Mercator's map

By treating the Earth as a cylinder rolled out flat, bearings could be plotted as straight lines.

- **1940s Aerial mapping** introduced. Photographs taken from aircraft covered large areas of the Earth. **Satellites*** are now used for mapping.

Taking satellite photographs ▼

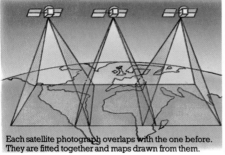

Each satellite photograph overlaps with the one before. They are fitted together and maps drawn from them.

Compasses

- **c.500BC** The **Greeks** used a naturally **magnetic* iron ore***, called **lodestone**, to make simple compasses.

- **c.AD1088 Shen Kua** (China) first mentioned a **magnetic needle**. He described a needle stuck through a straw floating in a bowl of water. Any magnetic needle that is free to swing will always point roughly north-south along the Earth's magnetic field, giving you an idea of your direction.

- **c.1187 A.Neckham** (GB) first referred to the **magnetic compass** in Europe. It was similar to the Chinese one.

Diagram of a simple compass ▼

True north

Magnetic north (the point at which the axis on which the Earth rotates passes through the planet).

Earth's magnetic field

- **By 1912 H.Anschütz-Kaempfe** (Germany) produced the first **gyrocompass**, based on the **gyroscope*** and used in ships. Once set to point north, a gyrocompass will remain stable, no matter how much the ship rolls.

- **1920s E.Sperry** (USA) introduced the **artificial horizon** for aircraft. This shows the angle of the aircraft to the ground.

Artificial horizon ▼

Turning to the left

Aircraft

Horizon bar

◀ The Pharos lighthouse

Lighthouses

- **c.285BC King Ptolemy II of Egypt** built the first known **lighthouse**, called the Pharos lighthouse, off Alexandria.

- **AD1780s Oil lamps*** were installed in lighthouses.

- **1862** The first **electric light*** was installed in a lighthouse at Dungeness, Scotland.

Theodolites and sextants

- 1st century AD Hero of Alexandria (Egypt) first described a **dioptra**, an early form of theodolite, used to plot the position of a site on a map.

- 1731 J.Hadley (GB) introduced an **octant**, an early form of sextant. A sextant is used to measure latitude, the angle between the horizon and the sun that shows how far north or south you are.

- c.1758 J.Campbell (GB) produced the first **sextant**, which is still the basic tool of navigation.

Early sextant ▼

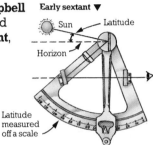

Latitude measured off a scale

- 1820s The **transit theodolite** was introduced in Europe. It was used to measure vertical and horizontal angles and distances of a site from other, known positions. It was the ancestor of the modern theodolite.

Modern theodolite ▶

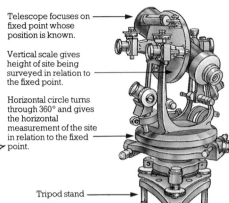

Telescope focuses on fixed point whose position is known.

Vertical scale gives height of site being surveyed in relation to the fixed point.

Horizontal circle turns through 360° and gives the horizontal measurement of the site in relation to the fixed point.

Tripod stand

Seismographs and geiger counters

- 1855 L.Palmieri (Italy) invented the first **seismometer**, an instrument for recording the strength and duration of an earthquake.

- 1880s J.Milne (GB) developed the **pendulum seismograph**, which is the basis of the modern seismograph.

Recording ▶ an earth tremor with a seismograph

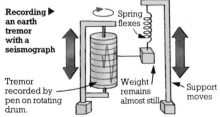

Spring flexes

Tremor recorded by pen on rotating drum.

Weight remains almost still

Support moves

- 1925-28 H.Geiger and W.Müller (Germany) developed the first **modern geiger counter**, a machine for detecting **radioactivity***. Radioactive materials send out invisible particles called **alpha particles***. Their strength can be measured on a meter or heard as clicks in earphones.

Sonar and radar

- By 1918 P.Langevin (France) and team had developed the **sonar** (sound navigation and ranging) system. It can be installed in submarines and ships to detect underwater objects by echo.

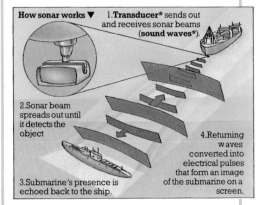

How sonar works ▼

1. **Transducer*** sends out and receives sonar beams (**sound waves***).

2. Sonar beam spreads out until it detects the object

3. Submarine's presence is echoed back to the ship.

4. Returning waves converted into electrical pulses that form an image of the submarine on a screen.

- 1931 W.Butement and P.Pollard (GB) built the first experimental **radar set**. The distance of ships from the shore was measured by timing how long **radio waves*** took to bounce back.

Alpha particles, 15; *Electric light*, 62; *Gyroscope*, 20; *Iron ore*, 22; *Magnetism*, 7; *Oil lamp*, 62; *Radioactivity*, 15; *Radio waves*, 53; *Satellite*, 38; *Sound waves*, 53; *Transducer*, 118.

73

Meteorology

The study of the weather is called meteorology and forecasters are called meteorologists. A variety of instruments have been developed to help forecast weather conditions.

Thermometers

- **c.1593 Galileo Galilei** (Italy) designed an instrument called a **thermoscope** to measure the temperature of air.

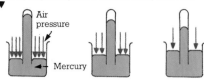

Galileo's thermoscope ▶

Glass tube

Coloured water

As air in the glass tube cooled, it shrank. Coloured water from the basin moved up the tube to fill the space. So, the higher the level of the liquid the cooler the temperature.

- **1615 G.Sagredo** (Italy) introduced the first **scale** for thermometers.

- By **1654 Grand Duke Ferdinand of Tuscany** (Italy) had made the first **sealed thermometer**. This excluded atmospheric pressure and so enabled more accurate readings to be made.

- **1718 D.Fahrenheit** (Germany) introduced the **fahrenheit scale** for measuring temperature.

- **1742 A.Celsius** (Sweden) introduced the **centigrade scale**.

Modern thermometers are marked with either ▶ fahrenheit or centigrade scales, sometimes both.

Barometers

- **1643 E.Torricelli** (Italy) made the first **barometer**, for measuring atmospheric pressure (the pressure of air around us).

In **Torricelli's barometer**, mercury was pushed up the tube by air pushing down on the mercury in a bowl.
▼

Air pressure

Mercury

- **1843 L.Vidie** (France) made the first **aneroid barometer**. (Aneroid means 'no liquid'.)

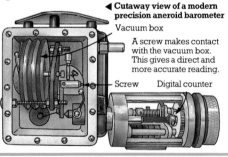

Vidie's aneroid barometer ▶

Pointer

Vacuum

A thin metal box with a **vacuum*** inside expanded or contracted according to changes in atmospheric pressure. This was then recorded by a pointer.

- **1960 J.Gradidge** (GB) patented the first **precision aneroid barometer**.

◀ Cutaway view of a modern precision aneroid barometer

Vacuum box

A screw makes contact with the vacuum box. This gives a direct and more accurate reading.

Screw   Digital counter

Barographs

- **1765 A.Cumming** (GB) made the first **barograph.** This draws a line on a revolving drum, giving a continual record of atmospheric pressure.

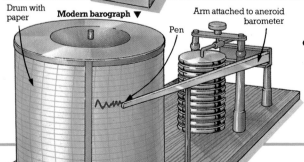

Drum with paper   **Modern barograph ▼**

Arm attached to aneroid barometer

Pen

Hygrometers

- **15th century Cardinal N.de Cusa** (Germany) designed the first **absorption hygrometer**, which measured humidity in the air.

- **1783 H.de Saussure** (Switzerland) made the first **hair hygrometer**. A length of human hair was attached to a pointer on a scale. Hair gets longer or shorter according to humidity and this was shown by the pointer.

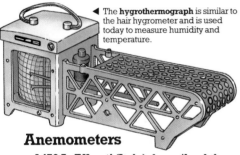

◀ The **hygrothermograph** is similar to the hair hygrometer and is used today to measure humidity and temperature.

Anemometers

- **c.1450 L.Alberti** (Italy) described the first **anemometer**, an instrument for measuring wind speed and direction. A weather vane had a small swinging plate. The stronger the wind, the greater the deflection of the plate.

- **c.1845 T.Robinson** (GB) invented the **cup anemometer**. Four cups rotated on a spindle.

- **1926 J.Patterson** (GB) improved efficiency by using **three cups** instead of four.

Modern three-cup anemometer ▶

Each rotation is recorded on a digital counter.

Amenograph ▼

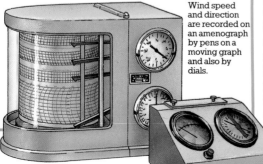

Wind speed and direction are recorded on an amenograph by pens on a moving graph and also by dials.

Rain-gauges

- **c.400BC** An early **rain-gauge** was described in an Indian manuscript. It was a bowl that measured the amount of rain that had fallen in a set period.

- **AD1722 Rev.Horsley** (GB) designed the first **modern rain-gauge**.

Modern rain gauge ▶

A funnel with a narrow tube prevents the rain already collected from evaporating.

Measuring glass

Rain collects in here

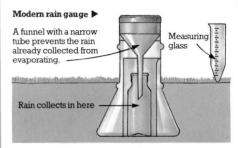

Measurements in the atmosphere

- **1749 A.Wilson** (GB) used **kites** to take temperature readings in the air. Thermometers were attached to the kite.

- **1930 P.Molchanov** (USSR) sent up the first **radio-sonde**. A small **radio*** in a balloon sent back readings of pressure, temperature, humidity and wind to a receiving station.

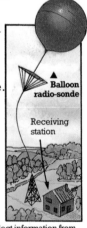

Balloon radio-sonde

Receiving station

Today, meteorological **rockets*** collect information from the earth's atmosphere and transmit it back to Earth.

- **1960 Tiros I** (USA) was the first **satellite*** to provide useful weather information. Orbiting satellites now scan the Earth, providing photographs and television pictures of the weather.

▲ Hurricane Gladys photographed from Apollo 7

Radio, 53; Rocket, 36; Satellite, 38; Vacuum, 118.

Optics

Instruments that help us to see are called 'optical', from the Greek word for eye. In the 1st century AD, the Roman writer Seneca noted that a glass sphere filled with water produced a magnifying effect. Spectacles were among the first and simplest optical inventions, but today there are many specialized optical instruments.

Lenses

- **11th century Ibn-al-Haitham** (Iraq) gave the first good description of a **lens**.

- **1267 Roger Bacon** (GB) described his experiments with lenses. These would have been hand-held **magnifying glasses**.

Magnifying glass ▶

Spectacles

- **c.1280 Spectacles** with **convex lenses** were first used in Venice, Italy. They were held in the hand. Later versions rested on the nose.

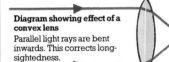

Diagram showing effect of a convex lens
Parallel light rays are bent inwards. This corrects long-sightedness.

- **1568 Concave lenses**, for correcting short-sightedness, were first mentioned in Italy.

Diagram showing effect of a concave lens
Parallel light rays are bent outwards.

- **1775 Benjamin Franklin** (USA) had the first **bi-focals** made for him. They combined both types of lens in one frame.

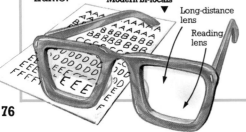

Modern bi-focals
▼ Long-distance lens
Reading lens

Contact lenses

- **1845 J.Herschel** (GB) first suggested the **idea of contact lenses**, small glass lenses which could be laid directly on the eye.

- **1887 E.Frick** (Switzerland) produced the first **contact lenses of sufficient precision** to be worn.

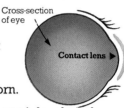

Cross-section of eye

Contact lens ▶

- **1932 J.Dallos** (Hungary) developed **individually fitting lenses** by taking moulds of the eyes.

- **1938 Orbig Co.** (USA) produced the first **plastic contact lenses**.

Binoculars

- **c.1608 H.Lippershey** (Holland) demonstrated the first **binoculars**. They consisted of a **telescope*** with two eyepieces.

- **19th century Prism binoculars** were developed. A prism is a piece of glass which can bend or reflect light.

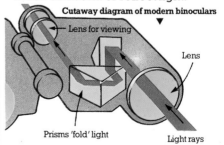

Cutaway diagram of modern binoculars
▼
Lens for viewing
Lens
Prisms 'fold' light
Light rays

The light to each eye is 'folded' by being reflected by a pair of glass prisms, so that the magnifying power of the binoculars is equal to that of a much longer telescope.

Microscopes

- **c.1590 H.Janssen** (Holland) possibly made the first **microscope**, an instrument for magnifying very small, close objects. It used a number of lenses to focus the image of the object.

- **1665 R.Hooke** (GB) designed the prototype of the **modern light microscope**. The body was attached to a side pillar so that it could be tilted. It had a lens at each end and one in the middle.

Hooke's ▶ microscope

- **1683 A.van Leeuwenhoek** (Holland) made and used very **powerful and accurate lenses** in simple, hand-held microscopes.

Cutaway diagram of a modern light microscope

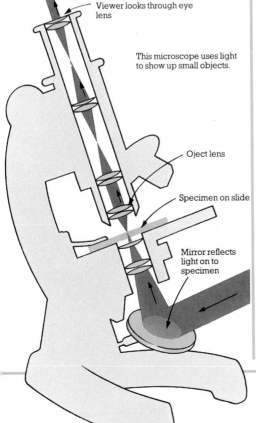

Viewer looks through eye lens

This microscope uses light to show up small objects.

Oject lens

Specimen on slide

Mirror reflects light on to specimen

Electron microscopes

- **1924 L.de Broglie** (France) discovered that beams of **electrons*** can move in waves like **light waves***, but with a much smaller wavelength. They can be used to magnify extremely small particles. A microscope using electron beams can magnify up to 1 million times.

- **1931 M.Knoll** and **E.Ruska** (Germany) built the first experimental **electron microscope**. The first commercial electron microscope was built in England in 1935.

Electron emitter

Electron beam

Magnet

Specimen

Magnets

Cutaway diagram of a modern electron microscope ▶

Electrons are emitted from heated metal and accelerated through a **vacuum***. They are then focused by powerful magnets on to the specimen. The magnified image appears on a screen.

Reflected electrons produce magnified image on screen.

To air pump which creates vacuum.

- **1938 M.von Ardenne** (Germany) built the first **scanning electron microscope**. A three-dimensional image of the specimen could be recorded on a photographic plate.

This picture, recorded by a ▶ **scanning electron microscope**, shows a cancer cell (centre) among the red blood cells of a leukemia sufferer. The scanning electron microscope has helped in the study of diseases.

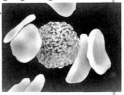

*Electron, 14; Light wave, see Laser, 21; Telescope, 78; Vacuum, 118.

Astronomy

People have been interested in astronomy, the study of stars and planets, since ancient times and a variety of different instruments have been developed for this purpose.

13th century AD brass Arabian astrolabe ▶

Pointers

The altitude of the Sun, Moon or stars could be read off the tables round the edge. This was done by turning the metal network so that the pointers lined up with the positions of the main stars, as the viewer saw them in the sky.

Moving network on backplate

Tables giving measurements

Astrolabes

- **c.150BC Hipparchus** (Greece) invented an **astrolabe**, an instrument used for plotting the altitude (the height above the horizon) of the Sun, Moon or stars. This was the main astronomical instrument used until the invention of the telescope.

Telescopes

- **c.1608 H.Lippershey** (Holland) designed the first **telescope**. It was called a **refracting telescope**.

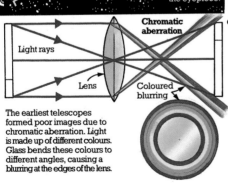

How a refracting telescope works ▼

Light rays

Eyepiece

Objective lens

Light from an object passes through and is focused by a lens called the objective. The image is then magnified by a second lens known as the eyepiece.

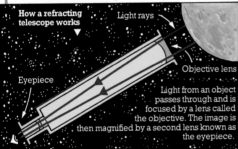

Chromatic aberration

Light rays

Lens

Coloured blurring

The earliest telescopes formed poor images due to chromatic aberration. Light is made up of different colours. Glass bends these colours to different angles, causing a blurring at the edges of the lens.

- **1663 J.Gregory** (GB) suggested using two **mirrors** to reflect the light. Mirrors reflect all light to the same angle, so that there is no chromatic aberration.

- **1668 Isaac Newton** (GB) built a **reflecting telescope** using mirrors. Giant modern telescopes are of the reflecting type.

Newton's reflecting telescope ▼

Light rays

Flat mirror

Eyepiece

Concave mirror

Light is collected by a **concave*** mirror and reflected on to a flat mirror at an angle of 45°. The light is sent to the side of the tube where the image is formed, focused and magnified.

- **1733 C.Hall** (GB) designed an **achromatic** (colourless) **refracting telescope**. Most smaller modern telescopes are based on this design.

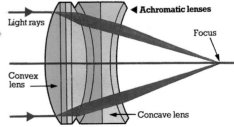

◀ **Achromatic lenses**

Light rays

Focus

Convex lens

Concave lens

An achromatic telescope contains a concave and a convex lens made of different types of glass. Each lens bends the light in opposite directions, cancelling out chromatic aberration.

- **c.1817 J.Fraunhofer** (Germany) introduced a **special mounting** to allow telescopes to move up and down and round.
- **1907 W.Gerrish** (USA) introduced **electric motordrive** on a telescope.

- **1969** The **largest reflecting telescope** in the world was built in northern Caucasus, USSR. It has a mirror measuring 600cm (236in) in diameter. This can cover huge areas of the sky.

Radio telescopes

- **1931 K.Jansky** (USA) first discovered that **radio waves*** can come from objects in outer space. He investigated radio waves coming from the constellation Sagittarius.

- **1937 G.Reber** (USA) built the first **radio telescope** with a dish reflector, 9.5m (31ft) across, to collect radio waves and focus them on to a receiver. Radio waves were shown as a line on a graph or made a noise in earphones.

- **1953** The **largest radio telescope** in the world was built at Arecibo, Puerto Rico.

Radio waves are reflected from a fixed dish, 305m (1,000ft) across, on to a receiver suspended overhead. Being fixed, it can only cover a narrow band of sky.

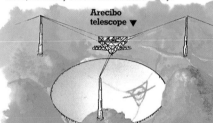

Arecibo telescope ▼

- **1957** The first large-scale, **movable radio telescope** was built at Jodrell Bank, England. It can cover a greater area of the sky than a fixed telescope.

Interferometers ▶
Some radio telescopes now consist of a number of antennae that move up and down rails so that they can cover even greater areas of the sky.

Spectroscopy

- **c.1665 Isaac Newton** (GB) produced the first experimental **spectrum**, demonstrating that white light is made up of a mixture of different colours.

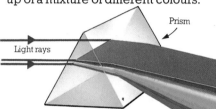

Prism

Light rays

Newton made his **spectrum** by passing ▲ light through a glass **prism***.

- **1814 J.Fraunhofer** (Germany) rediscovered **dark lines** across the Sun's spectrum, first recorded by **W.Wollaston** (GB) in 1802. They are now known as **Fraunhofer lines**. They were later shown to be caused by the absorption of light by different gases given off by the Sun.

- **1859 G.Kirchoff** and **R.Bunsen** (Germany) discovered that different substances give off different colours when burned. They built the first **spectroscope**, an instrument used to work out the chemical composition of planets by the light they emit.

A modern spectroscope ▼

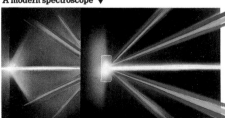

This uses a device called a diffraction grating to break up the light. This is a glass plate ruled with thousands of lines. The spectrum is photographed and analysed electronically.

* Concave mirror, see **Concave lens**, 76; **Prism**, see **Binoculars**, 76; **Radio waves**, 53.

Gravitation and the Universe

A force called gravitation brings together everything in the Universe. Gravitation keeps the planets in orbit round the Sun and holds the galaxies together. The

Earth's gravitation pulls down on every object on Earth with a force equal to the weight of that object. This force is also known as gravity.

Early ideas

- **c.AD150 Claudius Ptolemaeus** (Greece) published a book called the 'Almagest', in which he stated that **the Earth was the centre of the Universe**. This remained the official view until the 16th century.

◀ In the **Ptolemaic system**, the Earth lies at the centre of the Universe and all the other planets revolve round it.

- **1543 N.Copernicus** (Poland) published his theory that **the planets, including the Earth, all revolved round the Sun.**

The Copernican system ▶

- **1581 T.Brahe** (Denmark) built the first great observatory on the island of Hveen, off Denmark. He made very accurate **tables of planetry motion.**

- **1609 J.Kepler** (Germany) used Brahe's results to work out the basic **laws of the motion of the planets**. With these laws it became possible to draw up a scale map of the **Solar System***.

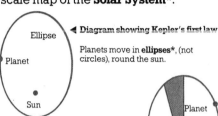

◀ **Diagram showing Kepler's first law**

Planets move in **ellipses***, (not circles), round the sun.

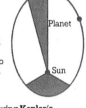

Diagram showing Kepler's second law ▶

The speed of the planets changes so that they move in equal areas round the sun (shaded, right) in equal periods of time.

◀ **Diagram showing Kepler's third law**

This provides a relationship between the time taken for a planet to complete one journey round the Sun and its distance from the Sun.

- **Early 17th century Galileo Galilei** (Italy) taught the new subject of **mechanics**, trying to work out how heavy things were, what was pushing and pulling them and how they moved.

Laws of motion

- **1687 Isaac Newton** (GB) published his **three laws of motion**. These laws are rules which relate force and motion. They explain the principles which govern the movement of all objects, on Earth and in space.

Newton's first law states that a body remains at rest, or in motion at a constant speed and in a straight line, except when this condition is changed by forces acting on it. This is the principle of inertia.

◀ **The Moon's orbit round the Earth**

According to Newton's first law, without the pull of the Earth's gravity, the Moon would move from A to B1. But because of the Earth's pull, the actual movement is from A to B2.

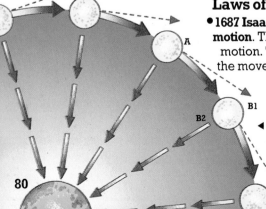

Newton's second law states that a force acting on an object causes it to accelerate. The size of the force needed to produce a given acceleration is equal to the mass of the object multiplied by acceleration. ▼

A given force on a given mass will produce a given acceleration.

On twice the mass, the same force will produce half the acceleration.

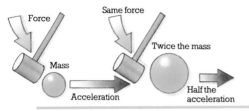

Force

Mass

Same force

Twice the mass

Acceleration

Half the acceleration

Newton's third law states that forces always occur between pairs of objects. If an object A exerts force on B (called the 'action'), then B exerts an equal and opposite force on A (called the 'reaction'). ▼

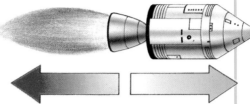

In order to take off, a **rocket*** ejects gas. The equal and opposite reaction of the gas on the rocket causes the rocket to move forward.

Theories of relativity

● **1905 Albert Einstein** (Germany) established his **special theory of relativity**. It is based on the idea that all motion is relative and there is no way of telling how fast we are moving through space. All we can measure is how fast we are moving in relation to some chosen 'frame of reference' (space and time).

Einstein's **special theory of relativity** states that all uniform motion is relative and that the speed of light is always constant. ▼

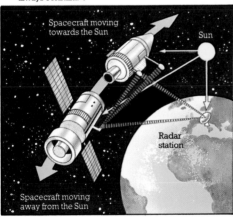

Spacecraft moving towards the Sun

Sun

Radar station

Spacecraft moving away from the Sun

Two spacecraft pass each other, each travelling at 16km (10 miles) per second as measured by a radar station on Earth.

But the pilots measure that they are travelling relative to each other at 32km (20 miles) per second.

If the pilots and the radar station measure the speed of light from the Sun, they all get the same result. The pilot moving towards the Sun does not get a measurement that reflects his movement relative to any other body.

● **1915 Albert Einstein** published his **general theory of relativity**. This was developed mainly to deal with problems of gravity.

Einstein's **general theory of relativity** states that behaviour in a gravitational field and an accelerating 'frame' are completely equivalent. ▼

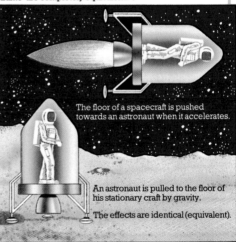

The floor of a spacecraft is pushed towards an astronaut when it accelerates.

An astronaut is pulled to the floor of his stationary craft by gravity.

The effects are identical (equivalent).

How light is bent by gravity ▼

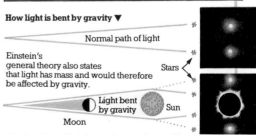

Normal path of light

Einstein's general theory also states that light has mass and would therefore be affected by gravity.

Stars

Light bent by gravity

Sun

Moon

The bending of light by gravity was detected in 1919 by photographing two stars in a **solar eclipse***. As the light rays from the stars passed the Sun, they were bent by its field of gravity. As a result, the two stars appeared to be further apart than usual.

Medicine: the fight against germs

Before the 19th century, many people in Europe died from diseases transmitted in unhygienic living conditions. Surgical operations were often fatal because lack of hygiene led to gangrene (decay of body tissue) and blood poisoning. During the 19th and 20th centuries, better methods were developed for treating illnesses.

Germs and disease

- **1676 A.van Leeuwenhoek** (Holland) first noticed **germs** (microbes such as bacteria) under his **microscope***. However, he did not realize their sigificance as a cause of disease.

Structure of a typical bacteria ▼

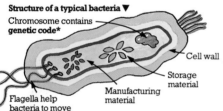

Chromosome contains **genetic code***

Cell wall

Storage material

Manufacturing material

Flagella help bacteria to move

- **1860 Louis Pasteur** (France) showed that the germs that fermented and soured wine could be destroyed by heating, without affecting the taste. This process is now called **pasteurization,** used for treating milk.

- **1865 Louis Pasteur** (France) published his **theory of germ disease**, claiming that bacteria can cause diseases.

Magnified view of bacteria ▼

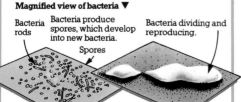

Bacteria rods

Bacteria produce spores, which develop into new bacteria.

Spores

Bacteria dividing and reproducing.

The air is filled with spores. If they settle on food or a wound, new bacteria will grow, causing infection.

Vaccination and immunization

- **1796 E.Jenner** (GB) prepared the first **effective vaccine** against smallpox. But Jenner was uncertain how it worked.

- **1885 Louis Pasteur** (France) prepared a **vaccine** against **rabies**. He showed that vaccines could be prepared from weakened germs, which would encourage a mild attack of the disease. This stimulated the body's defence system, making the person immune to later attacks.

- **1922 L.Calmette** and **C.Guérin** (France) developed a **tuberculosis vaccine**.

- **1954 J.Salk** (USA) prepared the first **poliomyelitis** (or **polio**) **vaccine**. Polio is a disease which killed and paralysed many people in the 1950s.

- **1960 J.Enders** (USA) developed a **measles vaccine**.

- **1970 Mass vaccination** was introduced in Europe and the USA.

How a vaccine works ▼

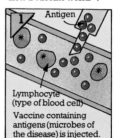

1 Antigen

Lymphocyte (type of blood cell)

Vaccine containing antigens (microbes of the disease) is injected.

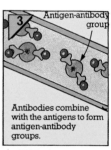

3 Antigen-antibody group

Antibodies combine with the antigens to form antigen-antibody groups.

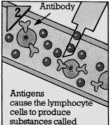

2 Antibody

Antigens cause the lymphocyte cells to produce substances called antibodies.

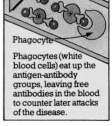

4

Phagocyte

Phagocytes (white blood cells) eat up the antigen-antibody groups, leaving free antibodies in the blood to counter later attacks of the disease.

Antisepsis

- 1867 **J.Lister** (GB) performed the first **antiseptic operation**, using a spray of carbolic acid to kill germs on the patient and in the air. **Antisepsis** is the technique of destroying germs already present.

Lister's carbolic spray ▼
Water boiler
Jet
Carbolic acid

Asepsis

- c.1880 **C.Chamberland** (France) designed the **autoclave**, a closed metal box that sterilized instruments and dressings by high-pressure steam. This is the first example of **asepsis**, the technique of preventing germs reaching operation wounds.

Modern ▶ electric autoclave

Modern methods of sterilization include ultra-violet light and **ultrasound***which break down the cell-structure of the microbes.

Chemotherapy

- 1891 **P.Ehrlich** (Germany) developed the concept of **chemotherapy**, the use of synthetic drugs to kill particular germs in body tissues.

- 1932 **G.Domagk** (Germany) developed a drug called **prontosil**, which kills the germs which cause diseases such as scarlet fever. It was the first of a class of drugs called **sulphonamides**, which can cure illnesses such as meningitis (a brain disease).

Chemotherapy is now also used to treat diseases not caused by germs, such as cancer.▼

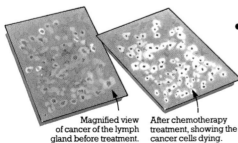

Magnified view of cancer of the lymph gland before treatment.
After chemotherapy treatment, showing the cancer cells dying.

Antibiotics

An **antibiotic** is a drug prepared from a living organism, often a mould, to kill other organisms which can cause illnesses such as pneumonia. An antibiotic is effective against bacteria but not viruses (diseases like measles).

- 1928 **Alexander Fleming** (GB) noticed that the mould **penicillium** stopped the spread of bacteria.

Penicillium destroying bacteria ▶
Strong bacteria
Healthy bacteria

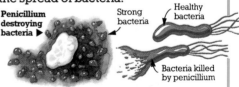

Bacteria killed by penicillium

- 1938–40 **H.Florey** (Australia) and **E.Chain** (Germany) isolated the active ingredient in the penicillium mould and called it **penicillin**.

Penicillium notatum ▶

The antibiotic is prepared from the rounded spores at the ends

Pain-killing drugs

- 1897 **F.Hoffmann** (Germany) produced the first synthetic **aspirin**. Aspirin, (acetysalicylic acid), is found naturally in certain plants and trees. It reduces aches and fever by lowering the body temperature. It also helps to ease rheumatism by reducing inflammation and swelling.

* Genetic code, 89; Microscope, 77; Ultrasound, 85.

Medicine: diagnosis

A doctor first has to find out what part of the patient's body an illness is in and what the illness is. This is called diagnosis. In the 19th century, instruments were designed which enabled internal organs to be studied without the patient being cut open.

Stethoscopes

- **1816 R.Laënnec** (France) designed the first **stethoscope**. It was called 'monaural' as it had one earpiece.

- **1850 G.Camman** (USA) designed the modern **binaural** (two ear) **stethoscope**.

◀ **Modern stethoscope**
A stethoscope works by amplifying the sounds made by the heart and lungs, providing information on their condition.

Ophthalmoscopes

- **1851 H.von Helmholtz** (Germany) designed the first **ophthalmoscope**, an instrument for seeing inside the eye. It consisted of an arrangement of mirrors which shone light through the pupil, to light up the retina at the back.

Modern ophthalmoscope ▼

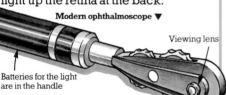

Viewing lens

Batteries for the light are in the handle

Endoscopes

- **c.1927 J.Baird** (GB) discovered the principle of **fibre optics***, in which light is conducted along strands of flexible glass. This made possible flexible inspection tubes for seeing inside the body.

- **1956 B.Hirschowitz** (S.Africa) first used a **fibrescope**, an endoscope that uses fibre optics. An endoscope is a long instrument for examining inside a body.

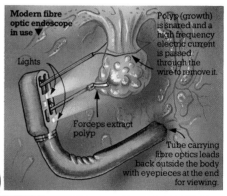

Modern fibre optic endoscope in use ▼

Lights

Polyp (growth) is snared and a high frequency electric current is passed through the wire to remove it.

Forceps extract polyp

Tube carrying fibre optics leads back outside the body with eyepieces at the end for viewing.

Thermometers

- **1625 S.Santorio** (Italy) first used a **thermometer*** to measure body temperature.

- **1863-64 W.Aitken** (GB) introduced the first **clinical thermometer**.

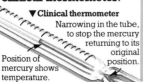

▼ **Clinical thermometer**
Narrowing in the tube, to stop the mercury returning to its original position.
Position of mercury shows temperature.

Sphygmomanometer

- **1896 S.Riva-Rocci** (Italy) designed the first practical **sphygmomanometer**, an instrument for measuring blood pressure. If blood pressure is too high or too low, it can affect the heart.

How a sphygmomanometer works ▶

A rubber cuff constricts the blood vessels, which stops the pulse. (The pulse is the expansion and contraction of an artery, caused by the beat of the heart). It is then released and the point at which the pulse returns indicates the blood pressure.

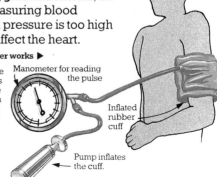

Manometer for reading the pulse

Inflated rubber cuff

Pump inflates the cuff.

Electrocardiographs

- 1903 W.Einthoven (Holland) designed the first accurate **electrocardiograph**, a device for measuring the regularity of the heartbeat. The heart's beat is caused by electric charges in the muscle fibre.

An **electrocardiograph** consists of a voltmeter, connected to the body by electrodes. ▼

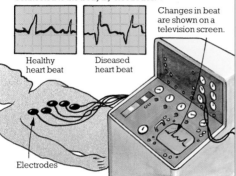

Changes in beat are shown on a television screen.

Healthy heart beat   Diseased heart beat

Electrodes

- 1929 H.Berger (Germany) designed the first **electroencephalograph**, which measures electrical activity in the brain. It can be used to detect tumours. Electrodes are placed on the head and electrical impulses relayed to a recorder.

X-rays

- 1895 W.Röntgen (Germany) first discovered **X-rays**. These are invisible waves of energy, like **light waves***. Unlike light, they can travel through things such as flesh, but are absorbed by denser materials like bone.

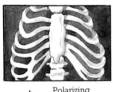

◄ **X-ray photograph of ribs**
The rays pass through the flesh, which comes out dark, but are absorbed by the bones, which show up lighter.

How an X-ray photograph is taken

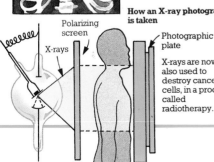

Polarizing screen

X-rays

Photographic plate

X-rays are now also used to destroy cancer cells, in a process called radiotherapy.

Ultrasound scanners

- 1955 I.Donald (GB) first successfully used **ultrasound**, high-frequency **sound waves***. He used it to study an unborn baby. This is preferable to X-rays, which can be harmful to unborn babies.

Ultrasound scanning ▼
Ultrasound waves are beamed on to a person's body. A complete picture of the body can be built up from the pattern of reflections given off by the sound waves.

The waves pass through the body and are reflected by solid objects, such as bone or muscle. A computer then reassembles the reflections into an image on a screen.

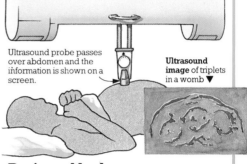

Ultrasound probe passes over abdomen and the information is shown on a screen.

Ultrasound image of triplets in a womb ▼

Brain and body scanners

- 1971 G.Hounsfield (GB) developed a **computerized axial tomographic (CAT) scanner**, a machine for seeing inside the brain.

- 1973 G.Hounsfield, **F.Doyle** and **L.Kreel** (GB) developed the first usable **body scanner**.

Using a scanner ▼

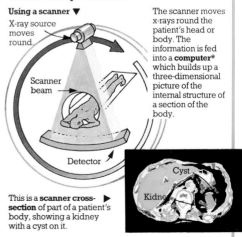

X-ray source moves round.

Scanner beam

Detector

The scanner moves x-rays round the patient's head or body. The information is fed into a **computer*** which builds up a three-dimensional picture of the internal structure of a section of the body.

This is a **scanner cross-section** of part of a patient's body, showing a kidney with a cyst on it. ►

Cyst

Kidney

*Computer, 49; Fibre optics, 52; Light waves, see Laser, 21; Sound waves, 53; Thermometer, 74.

Medicine: surgery and life support

Until the invention of anaesthetics in the 19th century, surgery was rarely successful, as the patients often died from shock.

Anaesthetics, drugs which cause a loss of feeling in all or part of the body, allowed new operations that could not have been performed on conscious patients.

Anaesthetics

- **c.1800 H.Davy** (GB) first noted the properties of **nitrous oxide gas** as a pain-killer.

- **1846 W.Morton** (USA) first successfully demonstrated the use of **ether** as an anaesthetic.

Ether flask ▶

Air was inhaled through a flask containing an ether-soaked sponge. A valve in the mouthpiece let out the expired air.

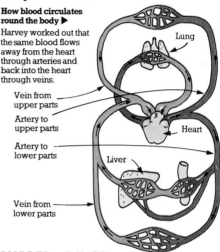

- **1847 J.Simpson** (GB) first used **chloroform**, to reduce pain at childbirth. This was quicker and more effective than ether.

◀ **Simpson's chloroform inhaler and container**

- **1854 S.Hardy** (GB) applied the first **local anaesthetic**, using chloroform in a **syringe***. Small areas of the body could be numbed and operations, such as dentistry, could now be performed without putting the patient to sleep.

The early forms of anaesthetic could be dangerous, as it was difficult to calculate how much of a drug to give a patient. With **modern anaesthetics**, pain and consciousness can be controlled much more precisely.

Blood transfusion

- **1625 W.Harvey** (GB) discovered the **circulation of blood** round the body. Before this, people believed that the blood passed backwards and forwards in two separate parts of the body.

How blood circulates round the body ▶

Harvey worked out that the same blood flows away from the heart through arteries and back into the heart through veins.

Vein from upper parts

Artery to upper parts

Artery to lower parts

Vein from lower parts

Lung

Heart

Liver

- **1825 J.Blundell** (GB) made the first successful **blood transfusion**, transferring blood from one person to another using a syringe.

- **1901 K.Landsteiner** (Austria) discovered the existence of **blood groups**. The blood group of the donor could now be matched to that of the patient.

Matching different bloods ▼

No reaction on mixing two blood groups.

Incompatible bloods mixed together form clots.

- **1914-15 L.Agote** (Argentina), **A.Hustin** (Belgium), **R.Lewisohn** and **R.Weil** (USA) discovered how to prevent blood clotting by adding sodium citrate. **Blood could now be stored**.

Kidney machines

- 1943 **W.Kolff** (Holland) designed the first practical **kidney machine**. This performs for the body the same tasks that the healthy kidneys would do.

Diagram of a kidney machine ▼

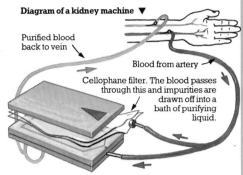

Purified blood back to vein

Blood from artery

Cellophane filter. The blood passes through this and impurities are drawn off into a bath of purifying liquid.

- 1960 **B.Scribner** (USA) designed a **plastic plug** which is fitted in a kidney patient's arm. This prevents damage caused by the frequent insertion of needles into the arm.

Cardiac pacemakers

- 1958 **A.Senning** (Sweden) implanted the first **internal cardiac pacemaker** into a patient's chest. This replaces the nerve which controls the electric charges producing heartbeats.

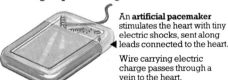

An **artificial pacemaker** stimulates the heart with tiny electric shocks, sent along leads connected to the heart.

Wire carrying electric charge passes through a vein to the heart.

Organ transplants

- 1943 **T.Gibson** and **P.Medawar** (GB) discovered the existence of the body's **defence system**. This destroys donor organs if they do not come from a near relative. Drugs were later developed to help reduce its effects.

- 1950 **R.Lawler** (USA) performed the first successful **kidney transplant** in which the patient survived.

Iron lungs

- 1929 **P.Drinker** and **C.McKhann** (USA) designed the first **iron lung**. This operates the lungs of patients suffering from diseases such as **polio*** which can paralyse lung muscles, causing suffocation.

Iron lung ▶

A pump creates a **vacuum*** in an airtight box.

Each time a vacuum is created in the box, the pressure of the atmosphere outside the box forces air into the patient's chest.

Heart-lung machines

- 1953 **J.Gibbon** (USA) performed the first open-heart operation using a **heart-lung machine**. The machine acts as the heart during the operation, circulating blood round the body and supplying oxygen to the lungs.

Diagram of a heart-lung machine ▼

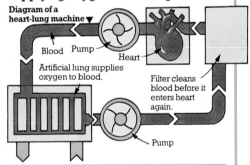

Blood Pump

Heart

Artificial lung supplies oxygen to blood.

Filter cleans blood before it enters heart again.

Pump

- 1967 **C.Barnard** (S.Africa) performed the first **heart transplant**. Today, heart transplant operations are quite common but the problems of rejection have not yet been overcome.

Heart transplant ▼

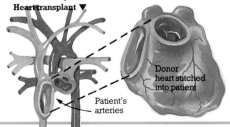

Donor heart stitched into patient

Patient's arteries

Chemistry

Since ancient times, people have been trying to find out what substances were made of and how they behaved when mixed with other substances. They tried to work out a pattern in that behaviour.

- **c.350BC Aristotle** (Greece) believed that everything was made up of combinations of just four elements: **air**, **earth**, **fire** and **water**. Aristotle laid the foundations of scientific study.

◄ The four elements according to Aristotle's theory

- **AD1661 R.Boyle** (GB) first defined the modern idea of an **element**: a substance that cannot be broken down into simpler forms.

- **1766 H.Cavendish** (GB) discovered the presence of a highly **inflammable*** gas, known today as **hydrogen**.

▼ Cavendish's experiments on factitious air

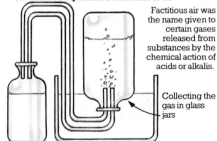

Factitious air was the name given to certain gases released from substances by the chemical action of acids or alkalis.

Collecting the gas in glass jars

- **1774 J.Priestley** (GB) first isolated the gas known today as **oxygen** by heating mercury.

- **1787 A.Lavoisier** (France) laid down the modern principle of **naming chemical compounds** after the elements they contain, in his book 'Methods of Chemical Nomenclature'. He also discovered that water was not a separate element but a chemical combination of hydrogen and oxygen.

- **1808 J.Dalton** (GB) published his **atomic theory***. He stated that elements could be divided into tiny particles called atoms and that atoms differed from each other by weight. He also explained chemical reactions, as the joining of atoms to form **molecules***.

This is **Dalton's list of symbols for atoms** of different elements, showing their atomic weights. ▼

- **1869 D.Mendeleyev** (USSR) published his **'Periodic Table of the Elements'**. This table grouped the elements according to their atomic structure and chemical properties. His work enabled chemists to identify 'families' of elements with similar chemical and physical properties. This is regarded as the backbone of chemistry.

Mendeleyev's periodic table of the elements ▼

The elements are arranged according to atomic weight. The elements side by side in adjacent columns behaved in the same way. The question marks represented elements which have been discovered since Mendeleyev.

			Ti=50	Zr=90	?=180	
			V=51	Nb=94	Ta=182	
			Cr=52	Mo=96	W=186	
			Mn=55	Rh=104	Pt=197	
			Fe=56	Bu=104	Ir=198	
			Ni=Co=59	Pl=106	Os=199	
H=1			Cu=63	Ag=108	Hg=200	
	Be=9	Mg=24	Zn=65	Cd=112		
	B=11	Al=27	?=68	Ur=116	Au=197?	
	C=12	Si=28	?=70	Sn=118		
	N=14	P=31	As=75	Sb=122	Bi=210	
	O=16	S=32	Se=79	Te=128?		
	F=19	Cl=35	Br=80	I=127		
Li=7	Na=23	K=39	Rb=85	Cs=133	Ti=204	
		Ca=40	Sr=87	Ba=137	Pb=207	
		?=45	Ce=92			
		?Er=56	La=94			
		?Yt=60	Di=95			
		?In=75	Th=118?			

* Atomic theory, 14; Inflammable, 118; Molecule, 118.

Biology

Biologists examine the bodies of plants and animals and try to understand how they work. They also look at the structures of the bodies and try to work out why they are the way they are. This has led to a number of theories and discoveries.

- **1753 Carl Linné** (Sweden), also known as **Linnaeus**, published his 'Species Plantarum', in which he devised a **system of classification**, giving each kind of plant two names.

◀ In **Linnaeus'** system, the first name given to a plant indicates its **genus**, or large family group to which it belongs. The second name gives the **species**, or specific type within the genus.

An orange tree is a *Citrus aurantium*.

A lemon tree is a *Citrus limon*.

- **1859 Charles Darwin** (GB) published his **theory of evolution by natural selection**. He argued that the plants and animals that are best adapted to the environment are the most likely to survive, breed and pass on their characteristics to their offspring. However, he did not know how the characteristics were passed on.

- **1866 G.Mendel** (Czechoslovakia) published his **law of inheritance**. This stated that offspring inherit one set of instructions (later known as 'genes') from each parent. Some instructions are 'dominant', others 'recessive'.

Diagram illustrating Mendel's experiment on sweet peas ▼

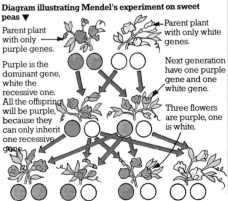

Parent plant with only purple genes.

Parent plant with only white genes.

Purple is the dominant gene, white the recessive one.

Next generation have one purple gene and one white gene.

All the offspring will be purple, because they can only inherit one recessive gene.

Three flowers are purple, one is white.

- **1920s J.Haldane, R.Fisher** (GB) and **S.Wright** (USA) worked out the **'synthetic' theory of evolution**. This drew together Darwin's and Mendel's ideas and demonstrated that if Mendel's laws were correct, Darwin's theory could account for the evolution of the variety of life in the world.

- **1944 O.Avery** (USA) demonstrated that it was a chemical called **DNA (deoxyribonucleic acid)** that carried the genetic message.

- **1953 Francis Crick** (GB) and **James Watson** (USA) discovered the **structure of DNA**. DNA is a chemical code of instructions which controls the way plants and animals look and the way their bodies work. The code transmits characteristics from one generation to the next.

Double helix structure of DNA ▶

The discovery of this structure explained how, when a cell divides to produce two identical new cells, it can pass characteristics from one generation to the next.

The two DNA strands separate to form two new strands each.

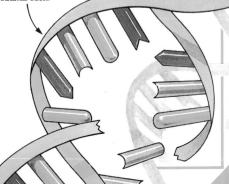

Warfare

The earliest weapons consisted of hand-held sticks, clubs, stone axes, knives and spears. Over the centuries, weapons have been made more destructive and able to cover greater distances. A missile today can travel over 12,875km (8,000 miles) at a speed of 24,000kmh (15,000mph).

Bows, arrows and swords

- **c.30,000BC** The oldest known pictures of **bows and arrows** appear in cave paintings in the Sahara Desert. The bow and arrow enabled people to kill beyond throwing range for the first time.

- **c.3500BC Swords** first appeared in the Middle East. They were made of **bronze*** cast in moulds.

- **c.1100BC Iron* swords** were made of red-hot iron, hammered into shape, producing harder and sharper blades.

- **500BC Sun Tzu** (China) first mentioned **crossbows** in a book called 'The Art of War'.

- **c.AD1100** The **European crossbow** was being used. The short bolt (arrow) could pierce **chain mail*** at a range of 365m (1,200ft).

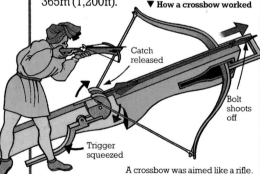

▼ How a crossbow worked

Catch released

Bolt shoots off

Trigger squeezed

A crossbow was aimed like a rifle.

- **c.1150** The **longbow** was developed in Wales. It could be fired very fast over a range of 228m (750ft).

Siege engines

- **c.500BC** The **Greeks** used **giant catapults** for shooting heavy missiles 365-457m (1,200-1,500ft).

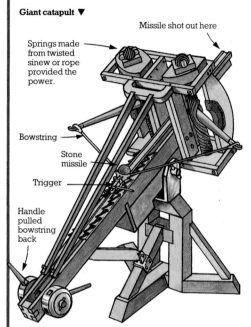

Giant catapult ▼

Missile shot out here

Springs made from twisted sinew or rope provided the power.

Bowstring

Stone missile

Trigger

Handle pulled bowstring back

- **c.AD1100-1400** A giant machine called a **trebuchet** was used in Europe to hurl large stones.

Explosives

- **c.1000** The **Chinese** were using **gunpowder** for fireworks and signals. Gunpowder contains saltpetre, charcoal and sulphur. It has to be ignited by flame or heat.

- **1846 A.Sobrero** (Italy) invented **nitroglycerine**, made from glycerol, nitric acid and sulphuric acid. It is extremely sensitive and can explode when hit.

- **1867 A.Nobel** (Sweden) made the first **dynamite**. It is a combination of nitroglycerine and materials such as wood pulp and sodium nitrate.

Cannon

- **c.1320** The **pot-de-fer** was the earliest form of gun. An arrow was shot from an iron pot using gunpowder.

- **c.1350-1850 Muzzle-loading cannon** were used in Europe. Until about 1450 they were made of **iron***, after that they were made of **bronze*** and iron.

Muzzle-loading ▼

1.Gunpowder pushed to bottom of barrel through the open end, or muzzle, followed by the missile.

2.Fuse leading to gunpowder was lit.

3.The gunpowder exploded, shooting the missile out of the barrel.

- **18th century   Lighter and more manoeuvrable cannon** were designed.

- **1850s   Breech-loading cannon** were introduced. This method was faster than muzzle-loading.

Breech-loading ▼

Missile and gunpowder inserted through door at bottom of barrel.

Door

Tight fit ensured more power and accuracy.

Modern M-107 cannon ▼

175mm gun fires shell over range of 32km (20 miles).

Vehicle has top speed of 54kmh (34mph).

Recoil spade gives firm hold on ground.

Guns

- **c.1350 Muzzle-loading handguns** were developed in Europe. The soldier lit a fuse (string) which he applied to the gunpowder in a small metal pan. This made a flash which set off the gunpowder in the barrel.

Matchlock mechanism ▼

Trigger   Fuse

When the trigger was squeezed a slow-burning match (fuse) was brought down on a pan of gunpowder. This set off the explosive charge.

- **Late 15th century** The **matchlock** mechanism was introduced in Europe. It enabled the soldier to keep both hands on the gun so that he didn't lose aim.

- **c.1610** The **flint-lock** mechanism was perfected in France. When the trigger was squeezed, a flint hit a steel plate, producing sparks which fell into the gunpowder pan and ignited it. This made shooting faster.

- **1718 J.Puckle** (GB) patented the first **machine gun**. When a handle was turned, bullets were brought round to a firing point in the barrel. The gun could fire 63 bullets in 7 minutes.

- **c.1805 J.Shaw** (USA) invented **percussion cap priming**. This made guns faster and more reliable.

Percussion cap priming ▶

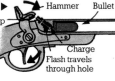

Hammer   Bullet

Cap

A hammer hit (like percussion) a metal cap containing explosives. The flash travelled through a hole, igniting the explosive charge.

Charge

Flash travels through hole

- **1812 S.Pauly** (Switzerland) invented the first **cartridge**. It had a cardboard body containing a bullet.

Firing pin strikes cap and detonates explosive charge.

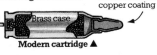

Lead bullet with copper coating

Brass case

Modern cartridge ▲

- **1879 J.Lee** (USA) introduced **magazine-loading**. Cartridges carried in a box below were pushed up into the rifle by a spring. This made firing extremely fast.

- **1884 H.Maxim** (USA/GB) patented the first **fully automatic machine gun**. The force of the recoil when the gun fired re-cocked it, so that it continued to fire automatically.

The modern **GE Vulcan** machine gun (USA) can fire 6,000 bullets in a minute. ▼

* Bronze, 22; Chain mail, 94; Iron, 22.

Warfare

Bombs and missiles

- **1849** The **Austrians** used the first **aerial bombs**, dropped on Venice in unmanned balloons.

- **1866 R.Whitehead** (GB) designed the first **torpedo**, an underwater missile launched from a ship. It was driven by compressed air and a **propeller***. Modern torpedoes use **sonar*** to find their targets.

- **1942 V-2 rocket*** (Germany), the first **surface-to-surface missile**, was launched. (This means launched from land to hit a land target.) It had a range of 320km (200 miles).

- **1944 V-I** (Germany), the first **jet*-powered flying bomb**, was developed. It had a range of 240km (150 miles).

V-I flying bomb ▼

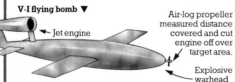

Jet engine

Air-log propeller measured distance covered and cut engine off over target area.

Explosive warhead

- **1945 J.Oppenheimer** (USA) led a research team that tested the first **atomic bomb**, a bomb whose power comes from **nuclear fission***.

- **1945 'Little Boy'** (USA) was the first **atomic bomb used in war**, dropped on Hiroshima, Japan.

Stages in the explosion of a 15 megatonne atomic ▼ bomb

1. Bomb explodes just above ground, forming a fireball.

2. High-pressure shock wave blows out from fireball. Buildings within 8km (5 miles) completely destroyed.

3. Fireball rises, sucking up dust and rubble.

4. This forms a cloud. Deadly radiation falls up to 300km (186 miles) away from the blast.

- **1952** The **hydrogen bomb** was first exploded by the USA. Its explosive power is based on **nuclear fusion***.

- **1957 SS-6** (USSR) was the first **intercontinental ballistic missile** (or **ICBM**), a missile designed for carrying nuclear warheads. It had a range of 9,600km (6,000 miles).

- **1970s** The **cruise missile** (USA), the **latest flying bomb**, was developed. It can be launched from land, sea (submarine) or air. It has a range of 3,200km (2,000 miles) and can strike within 30.5m (100ft) of its target.

Tomahawk cruise missile for land attack ▼

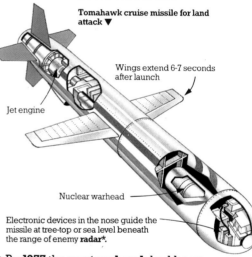

Wings extend 6-7 seconds after launch

Jet engine

Nuclear warhead

Electronic devices in the nose guide the missile at tree-top or sea level beneath the range of enemy **radar***.

- By **1977** the **neutron bomb** had been developed by the USA, although not put into production. In a neutron explosion more radioactive neutrons would be released than in a hydrogen explosion, killing more people but destroying less buildings.

- **1985** The first **anti-satellite* missile** was tested by the USA. The missile measured 5.5m (18ft) and was launched from a jet fighter. It contained no explosives as it destroyed its target by impact.

Battleships

- **c.2000BC** The **Egyptians** used **battleships** with oars and sails.
- **6th century BC** The **Greeks** used galleys called **triremes**. These had three layers of oars. Galleys were the main form of battleship until the 16th century AD.

Greek trireme (c.480BC) ▼

- **c.AD1500** The design of ships was changed to carry heavy **cannon***. They had more sails, no oars and were more manoeuvrable.

▲ **English galleon** (c.1585)

- **1860 HMS Warrior** (GB) was launched. It was the first **all-iron* battleship**. It had **steam engines*** as well as sails.
- **1905 HMS Dreadnought** (GB) was launched. It was the ancestor of the **modern battleship** and was driven by **steam turbines***.

▼ **HMS Dreadnought**

- **1918 HMS Argus** (GB) was the first **aircraft carrier**. It had a 168m (550ft) long flight deck and carried 20 aircraft.
- **1961 USS Enterprise** (USA) was launched. It was the first **nuclear-powered* aircraft carrier**.

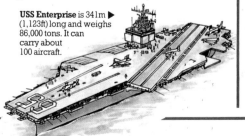
USS Enterprise is 341m ▶ (1,123ft) long and weighs 86,000 tons. It can carry about 100 aircraft.

Tanks

- **1916 Mark I** (GB), the first **tank**, was introduced. It was powered by an **internal combustion engine***.
- **1917 Renault FT17** (France) was the first tank of **modern design**, with a turret and gun mounted on top.

This is a modern Russian **T-72 tank**, introduced in 1974. It has a crew of three and maximum speed of 60km/h (43½mph). ▶

Bombers

- **1917 Gotha bombers** (Germany) first flew. They were the first aeroplanes to be designed as bombers. They were twin-engined **biplanes*** that cruised at 128kmh (80mph).
- **1980s Tupolev Blackjack** (USSR) was developed. This is the largest bomber in the world today. It has a maximum speed of 2,224kmh (1,382mph). **Tupolev Blackjack** ▲

Fighters

- **1915 Fokker E-I** (Germany) was the first **fighter plane**. It carried a **machine gun*** that fired through the propeller without hitting the blades.
- **1942 Messerschmitt ME262** (Germany) was the first **jet fighter** to fly. Two **jet engines*** gave it a speed of about 720kmh (550mph).
- **1980s Panavia Tornado** (GB/Italy/Germany) was introduced. It is one of the most advanced fighters.

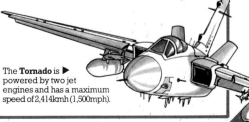
The **Tornado** is ▶ powered by two jet engines and has a maximum speed of 2,414kmh (1,500mph).

Radar, 95; **Satellite**, 38; **Sonar**, 73; **Steam engine**, 8; **Steam turbine**, 9; **V-2**, 36.

Warfare
Armour through the ages

- **c.2000BC** The **earliest known armour** appeared in Mesopotamia. It was made up of hundreds of small **bronze* scales**. This form of armour spread to the East and can be seen in Japanese samurai armour of the Middle Ages.

Suit of ▶ Japanese armour (c.AD1500)
Helmet and mask designed to frighten enemy

The armour was made up from steel scales. Coloured ribbons woven into the armour indicated the rank of the wearer.

- **6th century BC** The **Greeks** introduced the **bronze* cuirass** (breast plate) and **leg guards**.

◀ Greek armour

Cuirass had hinges at the shoulders and sides.

Leg guards pulled open and clipped round the legs.

- From **c.100BC Roman soldiers** wore armour made up from **iron bands** mounted on leather straps.

Iron bands mounted on leather

Roman ▼ armour
Iron helmet

Each cohort (unit) had a different coloured shield

Norman iron chain mail ▶ (c.AD1050)

- **c.AD1000** The **Normans** wore **chain mail** armour, made up from small interlocking metal rings.

- **14th century Steel plate armour** appeared in Europe. It was designed to cover all parts of the body.

Chain mail was worn underneath

Suit of steel plate armour (c.1400) ▶

English ▶ 'Roundhead' soldier (c.1650)
Steel breastplate

- By **1650** European armour consisted of **steel back-and-breast plates** only. This was because full armour thick enough to stop a bullet would have been too heavy to wear.

High leather boots

- **1914-18 Steel helmets** were re-introduced for infantry in Europe. **Gas masks** were introduced in **1915** following the first use of poison gas by the Germans.

British soldier with gas mask and steel helmet (1914-18) ▶
Box containing charcoal and lime-permanganate granules to filter the poisoned air.

Modern military clothing ▼

- **1980s Modern military protective clothing** consists of light jackets made of layers of very strong **nylon***.

94

Radar defence systems

● **1935 A. Wilkins** (GB) tested the first **radar* defence system. Radio waves*** detected an aeroplane 13km (8 miles) away at 3,050m (10,000ft). By 1939, Britain had a chain of radar stations round its southern and eastern coasts. They could detect enemy aircraft up to 160km (100 miles) away. Modern **anti-ballistic missile*** (ABM) systems have a range of up to 3,200km (2,000 miles).

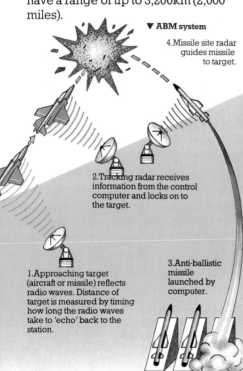

▼ **ABM system**

4. Missile site radar guides missile to target.

2. Tracking radar receives information from the control computer and locks on to the target.

1. Approaching target (aircraft or missile) reflects radio waves. Distance of target is measured by timing how long the radio waves take to 'echo' back to the station.

3. Anti-ballistic missile launched by computer.

● **1980s BAe Nimrod** (GB) carries an example of a modern **Airborne Early Warning (AEW) system.** Huge radar domes on the nose and tail of the aeroplane pick out every moving object above and below. The information is processed by **computer*** and sent to ground stations.

BAe Nimrod ▼

Radar dome
↓

Satellites and Star Wars

● **1980s Spy satellites*** have been developed which can take photographs of enemy military installations, giving information on missile sites, tanks and aeroplanes. The pictures are either beamed back to a receiver on Earth or ejected in a capsule.

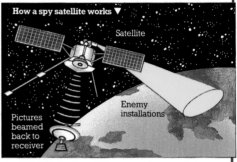

How a spy satellite works ▼

Satellite

Enemy installations

Pictures beamed back to receiver

● **1983 USA** announced plans for a **Strategic Defence Initiative (SDI)**, also known as 'Star Wars'. This is a plan to circle the world with a system of satellites that could shoot down any enemy **ICBMs*** with **laser beams***.

Possible arrangement for a ground-based SDI laser beam ▼

2. Relay mirror (37,014km/22,300 miles above Earth) deflects laser beam down on to battle mirror in lower orbit.

1. Laser station sends out beams to relay mirror.

3. Battle mirror aims laser beam at ICBM.

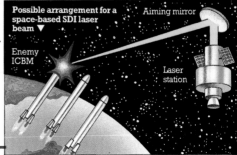

Possible arrangement for a space-based SDI laser beam ▼

Aiming mirror

Enemy ICBM

Laser station

Inventions and discoveries through the ages

All the entries listed below represent the earliest known examples of the invention, discovery or production of the item. The very earliest dates can only be approximate.

4,000,000-20,000BC

- **4,000,000 Tools**
  Stone choppers used in East Africa.

Early stone choppers from Tanzania, East Africa. ▲

- **500,000 Spears**
  Remains of sharpened wooden poles without spear heads found in England and Germany.

- **400,000 Buildings**
  Traces of huts made from branches found at Terra Amata, southern France.

- **250,000 Axes**
  Worked stone axes appeared in Africa, Asia and Europe.

- **50,000 Oil lamps**
  Remains of oil lamps, made from carved stone and using animal fat as fuel, found in Mesopotamia and Europe.

- **45,000 Spears**
  Stone spear heads attached to wooden shafts found in the Dordogne, France.

- **45,000 Paint**
  Cave paintings made with manganese oxide paints found in Europe, Middle East and Africa.

▲
Cave painting from southern Zimbabwe. It shows a kneeling figure of a person.

- **30,000 Bows and arrows**
  The earliest evidence found in cave paintings in the Sahara Desert.

- **20,000 Needles**
  Bone needles found in caves in the Dordogne, France.

20,000-5000BC

- **13,000 Harpoons**
  Barbed spears used for fishing found in the Dordogne, France.

- **12,000 Basketwork**
  Evidence for woven baskets found in the Middle East.

- **11,000 Nets**
  Nets used for fishing found in the Mediterranean area.

- **8000 Combs**
  Bone combs used in Scandinavia.

- **7500 Boats**
  The earliest evidence for water craft is a wooden paddle found at Star Carr, England.

- **7000 Pottery**
  Pottery vessels made in Iran.

- **6000 Beer**
  An alcoholic drink was made from grain in the Middle East.

- **6000 Sickle**
  A bone sickle set with flint teeth and used for cutting crops found in Jordan.

Egyptian wooden sickle set with flint teeth, c. 6000 BC.
▼

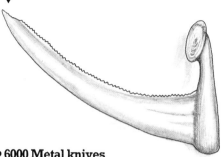

- **6000 Metal knives**
  Copper knives used in the Middle East.

- **6000 Bricks**
  Examples made of mud and dried in the sun found at Jericho, Jordan.

5000-2000BC

- **4500 Fish hooks**
  Examples found near the Black Sea at Lepinski Vir, Yugoslavia.

- **4000 Harnesses**
  Oxen were being harnessed for ploughing for the first time in Mesopotamia.

- **4000 Weighing instruments**
  Set of weighing scales found in Mesopotamia.

- **4000-3000 Drums**
  Remains of intruments made of animal skins stretched over frames found in the Upper Nile region of Africa.

- **4000-3000 Lathes**
  Wooden bow lathes used in the Middle East.

- **4000-3000 Sailing vessels**
  Sails on boats used on the Nile in Egypt.

- **3500 Writing**
  Cuneiform (wedge-shaped) writing began to develop in Mesopotamia.

- **3500 Bronze**
  Alloy of copper and tin was smelted in Mesopotamia.

- **3500 Nails**
  Copper nails used in Mesopotamia.

- **3500 Papyrus**
  A kind of paper made from layers of papyrus reeds beaten together used in Egypt.

- **3500 Potter's wheel**
  Wheel-made pottery appeared in Mesopotamia.

- **3500 Plough**
  Picture of a plough carved on a stone seal from Ur, Iraq.

- **3200 Wheel**
  A simple picture of a solid-wheeled vehicle found at Uruk, Iraq.

The Royal Standard of Ur, dating from 2750BC, shows carts with solid wooden wheels. ▲

- **3200 Ink**
  Lamp-black (soot) mixed with egg-white or honey used as an ink by the Egyptians.

• 3000 Dam
The Saad el-Kafara dam, built across the Garawi River Valley in Egypt to create an artificial lake for irrigation.

• 3000 Harp
Illustration of a stringed instrument, resembling a harp, found on the side of a vase from Iraq.

• 3000 24 hour day
The day was first divided into 24 equal periods of time by the Babylonians, Mesopotamia.

• 3000 Abacus
First used by the Chinese or Babylonians.

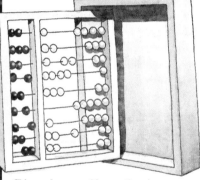

Chinese abacus used for counting. ▲

• 3000 Arch
Examples of arches at Ur, Iraq.

• 3000 Plumbline
This was a heavy weight used by Egyptian builders to plot vertical lines.

• 3000 Glass
Glass-making developed in Egypt and Mesopotamia.

• 3000 Cotton
Cultivated in the Indus Valley in Pakistan.

• 3000-2000 Pipes
Clay water pipes, used as drains, have been found at Mohenjo-Daro in Pakistan and at Knossos in Crete.

• 2800 Rope
Made from fibre of the hemp plant. Was used in China.

• 2600 Set square
Used by Egyptian stone masons.

• 2500 Skis
Found preserved in a bog at Hoting, Sweden.

Skis shown in a Swedish rock carving, c.2000BC. ▲

• 2500 Locks and keys
Wooden bolts closed by pins fitting into slots used by the Chinese.

• 2300 Map
The earliest surviving map is one inscribed on a baked clay tablet from Babylon, Iraq.

• 2000 Spoked wheels
Used on chariots in Mesopotamia.

• 2000 Bridles
Leather bridles for horses illustrated on Assyrian sculptures, Mesopotamia.

• 2000 Shadow clocks
Developed by the Egyptians.

2000-1000BC

• 1650 Swords
The earliest surviving examples found in shaft graves in Mycenae, Greece.

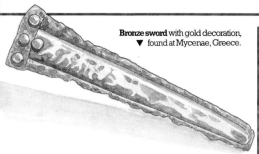

Bronze sword with gold decoration, ▼ found at Mycenae, Greece.

● **1500 Iron smelting**
Hittites from Anatolia, Turkey, introduced the technique to Europe.

● **1350 Welding**
A head-rest found in Tutankhamun's tomb in Egypt, but possibly made in Syria, carries the earliest known example of welding.

● **1300 Alphabet**
The earliest true alphabet, with letters representing simple sounds, appeared at Ugarit, Syria.

● **1200 Enamel**
Coloured glass paint used by the Egyptians for decorating mummy cases. It was applied as a liquid.

● **1000 Kites**
First appeared in China.

1000-500BC

● **800-700 Coinage**
King Gyges of Lydia, in Turkey, issued coins made from electrum, an alloy of gold and silver.

● **700 False teeth**
Carved bone or ivory dentures with gold braces worn by the Etruscans of northern Italy.

● **600 Cast iron**
First produced by the Chinese.

● **530 Calendar**
An accurate calendar, with months based on cycles of the Moon and the year based on cycles of the Sun, was in use in Babylon.

● **500 Magnet**
The magnetic properties of a type of iron called lodestone was first described by Thales of Miletus (Greece).

A piece of **magnetic lodestone**, wrapped in copper. ▲

● **500 Crossbow**
Sun Tzu of China mentioned a crossbow in a book called 'The Art of War'.

● **500 Iron share**
Appeared in Europe and made ploughing much faster.

● **500 Pens**
Quill pens were introduced in Europe and the Middle East.

500-0BC

● **500-400 Carpet**
A woven carpet, of Chinese or Iranian origin, was found in the Altai Mountains on the Chinese-Mongolian borders.

● **400 Pulley**
An early pulley used for raising loads is attributed to Archytas of Tarentum, southern Italy.

- **400-300 Shorthand**
  A system of writing notes quickly was devised by Xenophon of Greece.

- **300-250 Water pump**
  A plunger type instrument is attributed to Ctesibius of Alexandria, Egypt.

- **285 Lighthouse**
  The earliest known example built on the island of Pharos off the coast at Alexandria, Egypt.

- **236 Archimedean screw**
  Attributed to Archimedes of Syracuse, it was used to remove water from irrigation ditches in fields.

- **230 Springs**
  Designed by Ctesibius of Alexandria, Egypt and made from bronze plates. They were used in a catapult.

- **224 Valves**
  The earliest known were used by Ctesibius of Alexandria in a pump.

- **190 Parchment**
  Cured animal skins used as writing surfaces began to replace papyrus in Egypt.

- **150 Screw press**
  Machines were used by the Greeks to extract oil from olives and juice from grapes. A handle was turned to lower a beam on to the fruit.

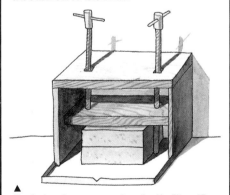

▲
The design of a **screw press** described by Pliny of Rome.

- **100 Central heating**
  Underfloor heating through hypocausts introduced by the Romans in Europe.

- **10 Crane**
  Mentioned by Vitruvius of Rome.

AD 0-500

- **79 Compasses**
  Bronze drawing compasses discovered at Pompeii, southern Italy.

- **c.100 Paper**
  The first true paper was being made by the Chinese from rags beaten with wood pulp, straw and water.

- **c.100 Steam engine**
  Designed by Hero of Alexandria, Egypt.

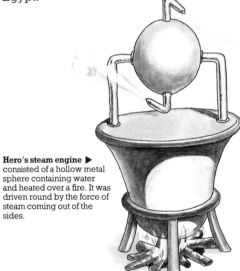

Hero's steam engine ▶ consisted of a hollow metal sphere containing water and heated over a fire. It was driven round by the force of steam coming out of the sides.

- **120-24 Dome**
  The first major use of this feature appeared in the Pantheon, Rome.

- **c.150 Ptolemaic system of the Universe**
  This stated that the Earth was the centre of the Universe and was devised by Claudius Ptolemaeus, Greece.

- **200 Skates**
  Iron skates were used in Scandinavia.

- **200-300 Wheelbarrow**
  Used in China for carrying heavy loads.

- **300 Stirrups**
  Stirrups were depicted on pottery tomb figures from China. Introduced into Europe in the 8th century.

- **362 Hospital**
  The first public hospital for the sick was opened in Rome by the Emperor Julian.

- **400 Decimal numbers**
  The first decimal system of numbers used in India

500-1000

- **500 Stencil**
  Used in China and Japan for reproducing text.

- **600 Chess**
  The earliest reference to chess first appeared in the Sanskrit romance 'Vasavadatta', India.

- **644 Windmill**
  The first windmill, with a vertical axis, is recorded in Iran.

- **700 Iron smelting**
  The Catalan Forge, an ancestor of the blast furnace, introduced in Spain.

- **800 Musical notation**
  Modern musical notation appeared in Europe.

- **800-900 Rudder**
  This was introduced by the Chinese. It gave better steering for ships.

- **c.813 Observatories**
  Astronomical observatories set up at Damascus, Syria and at Baghdad, Iraq.

- **834 Crank**
  An illustration of a crank appeared in the 'Utrecht Psalter', Holland.

- **851 Porcelain**
  Chinese porcelain drinking vessels were first described by the Arab traveller and scholar, Soleiman.

- **868 Printing**
  The earliest printed book, the Buddhist religious text 'The Diamond Sutra', was published in China by Wang Chieh.

- **900-1000 Plough**
  The wheeled plough was being used in Europe.

- **969 Playing cards**
  Recorded in China when Duke Ch'ien was murdered over a game of cards.

1000-1500

- **c.1000 Lens**
  The magnifying properties of a glass lens were first described by the Arab scientist Ibn-al-Haitham.

- **c.1000 Camera obscura**
  Developed by the Arabs.

- **c.1000 Gunpowder**
  Fireworks and signalling devices using a non-explosive type of gunpowder were being used by the Chinese.

Early Chinese arrow-rockets ▼

- **1040-50 Movable type**
  Invented by Pi Sheng, China.

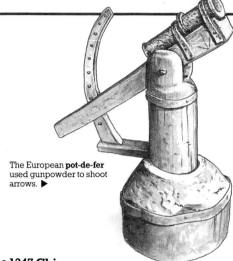

The European **pot-de-fer** used gunpowder to shoot arrows. ▶

- **1065 Stained glass**
The earliest known stained glass was fitted in some of the windows of Augsburg Cathedral, Germany.

- **1088 Clock**
A water-powered mechanical clock was made by Han Kung-Lien of China.

- **c.1088 Compass**
A simple form of compass, with a magnetic needle floating in water, was described by Shen Kua, China.

- **1150 Longbow**
Developed in Wales.

The Welsh longbow was about the same height as the archer. It was much faster to load than other bows and could be fired over longer distances. ▶

- **1250 Screw jack**
Designed by V.de Honnecourt, France.

- **1250 Tinplate**
Iron plated in a thin skin of tin was used in Czechoslovakia in the construction of a suit of armour.

- **1267 Magnifying glass**
Described by Roger Bacon, England.

- **c.1280 Spectacles**
The earliest record of spectacles appeared in S.di Popozo's 'Treatise on the Conduct of the Family', Venice, Italy.

- **1320 Lace**
First made in France and Belgium.

- **1320 Cannon**
The pot-de-fer, the earliest form of cannon, appeared in Europe.

- **1347 Chimneys**
The earliest reference to chimneys appears in Venice, Italy, recording the destruction of chimneys in an earthquake.

- **c.1350 Muzzle-loading guns**
Developed in Europe.

- **1405 Screws**
Metal screws were first mentioned in Kyeser's 'Bellafortis'.

- **1410 Coil springs**
First used in a clock designed by F.Brunelleschi, Italy.

- **1415 Oil paints**
Introduced in painting by Jan and Hubert van Eyck, Belgium.

- **1421 Explosive shells**
Used at the Siege of St.Boniface, Corsica.

- **1430 Flywheel**
Used in Germany to produce a smooth flow of power from a crank and connecting rod.

- **c.1450 Printing press**
Invented by Johannes Gutenberg, Germany. He produced the first European printed book called the 'Donatus Latin Grammar'.

- **1489 Arithmetic: plus (+) and minus (-) signs**
  Used in a book called 'Mercantile Arithmetic' written by J.Widman, Germany.

- **1495 Whisky**
  Distilled by J.Cor, Scotland.

1500-1600

- **1500 Caesarian operation**
  Believed to have been performed by J.Nufer, Switzerland, on his wife.

- **c.1500 Pocket watch**
  Made by P.Henlein, Germany.

- **1509 Wallpaper**
  Printed by H.Goes, England. It was a black and white imitation brocade.

- **1512 Theodolite**
  A 'polimetrum' was designed by M.Waldseemüller, France. It was the prototype of the theodolite, having the two essential devices for simultaneous measurement of horizontal and vertical angles.

- **1538 Diving bell**
  The earliest reference appears in Toledo, Spain.

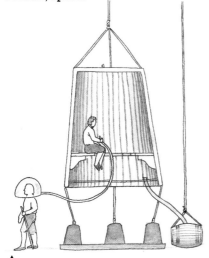

Design for a diving bell by Sir Edmund Halley, 1717.

- **1540 Artificial limbs**
  Fitted to wounded soldiers by A.Paré, France.

- **1543 Copernican theory of the Universe** Published by Nicolaus Copernicus, Poland. This stated that the Sun was the centre of the Universe.

- **1557 Arithmetic: equals sign (=)**
  First used by R.Recorde, England in his book, 'The Whetstone of Wit'.

- **1558 Encyclopaedia**
  The first book to be called an 'Encyclopaedia' written by P.Scalich, Switzerland.

- **1550s Muskets**
  Made in Spain.

- **1561 Dredger**
  The first machine designed for clearing canals was built by Pieter Breughel, Belgium.

- **1564 Contraceptive sheath**
  The first, made of linen, was designed by G.Fallopius, Italy.

- **1565 Pencil**
  The first pencil, with the lead made of pure graphite and with a wooden cover, was designed by K.Gesner, Switzerland. Modern pencil lead consists of graphite and clay.

Gesner's pencil ▲

- **1568 Lathe**
  A metal lathe for cutting screws was designed by J.Besson, France.

- **1569 Map with Mercator projection**
  Issued by Gerhard Kremer, known as Mercator, Holland.

1585 Naval mine
Floating mines, triggered by clockwork, were used by the Dutch at the Siege of Antwerp.

1589 Water closet
A lavatory with an underground sewage tank was installed by J.Harington, England.

c.1590 Microscope
Possibly made by H.Janssen, Holland.

c.1593 Thermometer
A glass tube with coloured water that was affected by air temperature was invented by Galileo Galilei, Italy.

1594 Grenades
Explosive hand grenades were being used in France.

1600-1700

c.1608 Telescope
Designed and demonstrated by Hans Lippershey, Holland.

This **telescope** was built by Galileo Galilei to the ▲ design of Hans Lippershey.

1609 Thermostat
A device for regulating the heat of stoves was invented by C.Drebble, Holland.

1609 The motion of planets
The three basic laws for the motion of planets were formulated by J.Kepler, Germany.

1609 Newspaper
The earliest known weekly newspaper was the 'Aviso Relation oder Zeitung', published by A.von Sohne, Germany.

1614 Logarithms
First worked out by J.Napier, Scotland.

1620s Submarine
A submarine made of greased leather over a wooden framework was demonstrated on the River Thames, London by C.Drebble, Holland.

1628 Circulation of the blood
Discovered by William Harvey, England, and illustrated in his book called 'De Motu Cordis'.

Arithmetic: multiplication sign (x)
First used by W.Oughtred, England.

1637 Umbrellas
Waterproof umbrellas were used by Louis XIII of France.

1642 Calculating machine
An adding and subtracting machine was invented by Blaise Pascal, France.

1643 Barometer
The principle of the barometer was demonstrated by E.Torricelli, Italy.

1650 Air pump
Demonstrated by O.von Guericke, Germany.

Guericke's air pump ▶

1654 Vacuum pump
Demonstrated by O.von Guericke.

1657 Pendulum clock
Designed by C.Huygens, Holland.

1661 Paper money
The first European bank-note was issued by the Bank of Stockholm, Sweden.

1661 Spirit level
Invented by M.Thévenot, France.

c.1666 Calculus
Developed by Isaac Newton, England.

1668 Reflecting telescope
Invented by Isaac Newton, England.

c.1670 Champagne
Produced by Dom Pierre Pérignon, France.

1674 Calculating machine
A multiplication and division machine was invented by G.Leibniz, Germany.

1675 Spring-driven clocks
Designed by C.Huygens, Holland.

1679 Pressure cooker
Invented by D.Papin, France.

Papin's pressure cooker ▶ was made of iron and was fitted with an air-tight lid. It allowed liquids to boil at higher temperatures than usual, cooking food in less time.

1687 Laws of Motion
The three Laws of Motion were published by Isaac Newton, England.

1698 Steam engine
A machine for draining flooded mines was patented by T.Savery, England. It was first manufactured in 1702.

1700-1800

1700 Clarinet
Invented by J.Denner, Germany.

c.1701 Seed drill
Invented by Jethro Tull, England.

1709 Iron smelting: coke-burning process
Introduced by Abraham Darby, England.

1709 Piano
Built by B.Cristofori, Italy.

1711 Tuning fork
Invented by J.Shore, England.

1712 Piston-operated steam engine
Built by T.Newcomen, England.

1716 Hot water central heating
Installed in a greenhouse in Newcastle upon Tyne, England by M.Triewald, Sweden.

1718 Machine-gun
The first rapid firing machine gun was patented by J.Puckle, England.

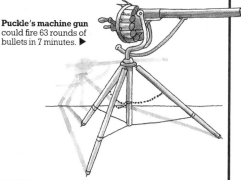

Puckle's machine gun could fire 63 rounds of bullets in 7 minutes. ▶

1718 Fahrenheit temperature scale
Devised by D.Fahrenheit, Germany.

1733 Flying shuttle
A device for speeding up cotton weaving was invented by J.Kay, England.

1734 Fire extinguisher
Invented by M.Fuches, Germany.

- **1742 Centigrade temperature scale**
First applied to a thermometer by A.Celsius, Sweden. Modern thermometers use either the Fahrenheit scale or the Centigrade scale, sometimes both.

- **1743 Silver plate**
The process of plating base metals with silver was invented by T.Boulsover, England.

- **1745 Wrought iron**
The first practical mass-production process for wrought iron was devised by C.Polhem, Sweden.

- **1748 Steel pens**
Made by J.Janssen, Germany.

- **1752 Lightning conductor**
Devised by Benjamin Franklin, USA.

- **c.1758 Sextant**
The first sextant, a modification of the octant, was designed by J.Campbell, England.

- **1759 Marine chronometer**
The first instrument sufficiently accurate for ocean navigation was made by J.Harrison, England.

▲
Harrison's marine chronometer was used to plot the longitude of a ship at sea.

- **1764 Spinning Jenny**
The first successful spinning machine was invented by J.Hargreaves, England.

- **1766 Fire escape**
A basket on a pulley and chains was patented by D.Marie, England.

- **1769 Steam engine**
A steam engine with a separate condenser was patented by James Watt, England.

- **1770 Steam carriage**
A steam-driven, three-wheeled gun carriage was designed by N.Cugnot, France.

- **1770 False teeth**
Dentures made from porcelain were introduced by A.Duchateau, France. They were the first really comfortable, close fitting dentures.

- **1777 Iron ship**
The first boat with an iron hull was built in Yorkshire, England.

- **1779 Spinning mule**
Invented by S.Crompton, England. It combined the features of the Spinning Jenny and Arkwright's water frame.

- **1783 Steam boat**
Built by the Marquis de Jouffroy d'Abbans and tested on the River Seine, France.

- **1783 Hot-air balloon**
Demonstrated by the Montgolfier brothers, France.

- **1783 Gas balloon**
Demonstrated by J.Charles, France. It was filled with hydrogen gas.

Charles' gas balloon ▶

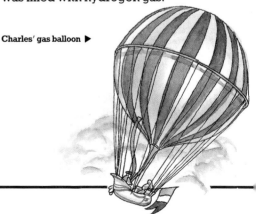

- **1784 Raised type**
  Introduced to help blind people to read more easily by V.Hany, France.

- **1786 Threshing machine**
  Invented by A.Meikle, Scotland.

- **1786 Clutch**
  Invented by J.Rennie, Scotland. It was used on factory machinery.

- **1787 Power loom**
  Patented by the Rev.E.Cartwright, England.

- **1791-95 Metric system**
  Officially adopted in France.

- **1792 Cotton gin**
  Invented by E.Whitney, USA. It enabled cotton to be cleaned 50 times faster than by hand.

- **1792 Gas lighting**
  A room was lit with coal gas lighting by W.Murdock, England.

- **1792 Ambulance**
  The first vehicle designed to carry wounded people was designed by D.Larrey, France.

Larrey's horse-drawn ambulance ▲

- **1795 Hydraulic press**
  Invented by J.Bramah, England.

- **1796 Vaccination**
  The first vaccine was given to prevent smallpox by E.Jenner, England.

- **1798 Lithography**
  Invented by A.Senefelder, Czechoslovakia.

- **1799 Gas fire**
  Patented by P.Lebon, France.

1800-1850

- **1800 Battery**
  Made by Alessandro Volta, Italy.

- **1803 Steam train**
  Built by R.Trevithick, England.

- **1807 Street lighting**
  Gas street lamps were installed in Pall Mall, London.

- **1808 Typewriter**
  Built by P.Turri, Italy.

- **c.1811 Canned food**
  Produced by Donkin and Hall, London.

- **1812 Hydraulic jack**
  Invented by J.Bramah, England.

- **1814 Spectroscope**
  Invented by J.Fraunhofer, Germany.

- **1815 Miner's safety lamp**
  Invented by Humphry Davy, England.

- **1816 Stethoscope**
  The monaural stethoscope was designed by R.Laënnec, France.

Laënnec's monaural stethoscope ▲

- **1818 Tunnelling machine**
  Designed by M.Brunel, England.

- **1820 Electromagnet**
  Built by H.Oersted, Denmark.

- **1822 Photograph**
  The first permanently fixed image was made by N.Niépce, France.

- **1823 Waterproof clothing**
  The first rain-proof coat was made by C.Macintosh, Scotland.

- **1824 Braille**
  A system of raised dots to help blind people to read was invented by L.Braille, France.

- **1827 Electricity**
  The law relating current, voltage and resistance was formulated by G.Ohm, Germany.

- **1827 Water turbine**
  The first practical water turbine was built by B.Fourneyron, France.

- **1830 Sewing machine**
  Invented by B.Thimonnier, France.

- **1831 Transformer**
  Invented by Michael Faraday, England.

- **1831 Dynamo**
  Demonstrated by Michael Faraday in London. The first one was built in 1832 by H.Pixii, France.

- **1832 Corrugated iron**
  Made by J.Walker, England.

- **1835 Photography: negative-positive process**
  Invented by H.Fox Talbot, England.

- **1839 Bicycle**
  Built by K.Macmillan, Scotland.

- **1839 Electric telegraph**
  The first regular service was fitted on the Great Western Railway, London.

  The first electric telegraph ▶

- **1840 Postage stamp**
  The 'Penny Black', the first postage stamp, was introduced by the British post office.

- **1846 Nitro-glycerine explosives**
  Made by A.Sobrero, Italy.

1850-1900

- **1850 Refrigeration plant**
  Set up by J.Harrison, Australia.

- **1851 Ophthalmoscope**
  Invented by H.von Helmholtz, Germany.

- **1851 Modular construction**
  The first large building of this kind was the Crystal Palace, London.

- **1852 Steam-powered airship**
  Built by H.Giffard, France.

- **1853 Hypodermic syringe**
  Invented by G.Pravaz, France.

19th century glass and metal syringe ▲

- **1856 Bessemer process**
  A method of mass-producing steel cheaply was perfected by H.Bessemer, England.

- **1856 Synthetic dye**
  The first synthetic dye, a purple dye called mauveine, was invented by W.Perkin, England.

- **1859 Evolution**
  The Theory of Evolution by Natural Selection was published by Charles Darwin, England.

- **1860 Pasteurization**
  The process of wine and milk pasteurization was worked out by Louis Pasteur, France.

- **1860 Internal combustion engine**
  The first, which ran on gas, was built by E.Lenoir, Belgium.

- **1862 Milking machine**
  The first machine was designed by L.Colvin, USA.

- **1862 Plastic**
  The first plastic was made from natural plant cellulose by A.Parkes, England.

- **1863 Dental drill**
  A clockwork drill was patented by G.Harrington, England.

Early clockwork dental drill ▶

- **1866 Law of inheritance**
  Published by G.Mendel, Czechoslovakia.

- **1867 Antisepsis**
  The first antiseptic operation was performed by J.Lister, Scotland.

- **1867 Reinforced concrete**
  Patented by J.Monier, France.

- **1867 Dynamite**
  Patented by A.Nobel, Sweden.

- **1869 Air brakes**
  Patented by G.Westinghouse, USA.

- **1874 Modern typewriter**
  Produced by C.Sholes and G.Glidden, USA.

Early Sholes and Glidden typewriter ▶

- **1876 Telephone**
  Patented by Alexander Graham Bell, USA.

- **1877 Record player**
  The phonograph was introduced by Thomas Edison, USA.

- **1879 Electric light**
  The first electric light bulb was invented by Thomas Edison, USA.

- **1884 Steam turbine**
  The first successful steam turbine was invented by C.Parsons, Ireland.

- **1884 Rayon**
  This was the first artificial fibre, produced by H.de Chardonnet, France.

- **1885 Motorcycle**
  Introduced by Gottlieb Daimler, Germany.

- **1885 Motor car**
  The first petrol-driven car was completed by Karl Benz, Germany.

- **1886 Linotype printing machine**
  Patented by O.Mergenthaler, Germany.

- **1886 Pre-stressed concrete**
  Invented by C.Dochring, Germany.

- **1887 Monotype printing machine**
  Invented by T.Lanston, USA.

- **1887/88 Radio waves**
  First proved to exist by H.Hertz, Germany.

- **1888 Pneumatic bicycle tyre**
  Designed by J.Dunlop, Northern Ireland.

- **1891 Chemotherapy**
  Developed by P.Ehrlich, Germany.

- **1894 Escalator**
  The Reno Inclined Elevator was designed by J.Reno, USA.

- **1894 Radio waves for communication**
Demonstrated by Guglielmo Marconi, Italy.

- **1895 X-rays**
Discovered by W.Röntgen, Germany.

- **1896 Radioactivity**
Discovered by A.Becquerel, France.

- **1897 Electrons**
Discovered by J.Thomson, England.

1900-1950

- **1900 Quantum Theory**
Formulated by Max Planck, Germany.

- **1901 Electric hearing aid**
Patented by M.Hutchinson, USA.

- **1901 Blood groups**
A,B and O blood groups were worked out by K.Landsteiner, Austria.

- **1901 Radio message**
Transmitted across the Atlantic by Guglielmo Marconi, Italy.

- **1903 Electrocardiograph**
Invented by W.Einthoven, Holland.

- **1903 Aeroplane flight**
The first controlled, powered flight was made by O.Wright, USA.

- **1904 Photoelectric cell**
Developed by A.Korn, Germany, for scanning photographs to be transmitted by wire.

- **1904 Diode valve**
Amplifying valve, patented by J.Fleming, England.

Fleming's diode valve was used to amplify weak ▶ electrical signals.

- **1905 Special Theory of Relativity**
Formulated by Albert Einstein, Germany.

- **1906 Triode valve**
Amplifying valve, patented by L.de Forest, USA.

- **1907 The 'plum pudding' model of the atom**
Proposed by J.Thomson, England.

- **1911 The nuclear model of the atom**
Proposed by E.Rutherford, New Zealand.

- **1913 Stainless steel**
Produced by H.Brearley, England.

- **1913 The structure of the atom based on quantum physics**
Proposed by N.Bohr, Denmark.

- **1915 General Theory of Relativity**
Published by Albert Einstein, Germany.

- **1919 Proton**
Identified by E.Rutherford, New Zealand.

- **1921 Motorway**
The first was the Avus Autobahn, Germany.

- **1921 Insulin**
Insulin, used in the treatment of diabetes, was first isolated by F.Banting and C.Best, Canada.

- **1922 Motion pictures**
The first commercially produced 'talkie' was 'Der Brandstifter' ('The Arsonist'), Germany.

- **1925 Television**
The first demonstration of television was given by J.Baird, Scotland.

- **1926 Liquid-fuel rocket**
Launched by R.Goddard, USA.

- **1928 Antibiotic**
Penicillin was discovered by Alexander Fleming, Scotland.

- **1930 Jet engine**
  Patented by Frank Whittle, England.

- **1935 Fluorescent lighting**
  Demonstrated by the General Electric Co., USA.

- **1935 Nylon**
  Patented by W.Carothers, USA.

- **1937 Radio telescope**
  Built by G.Reber, USA.

- **1939 Nuclear fission**
  Discovered by O.Hahn, Germany.

- **1942 Nuclear reactor**
  The first nuclear chain reaction was set up by E.Fermi, Italy.

- **1942 Aqualung**
  The first aqualung, for deep-sea diving, was designed by J.Cousteau, France.

- **1943 Kidney machine**
  Built by W.Kolff, Holland.

- **1943 The body's immune system**
  Discovered by T.Gibson and P.Medawar, England.

- **1945 Atomic bomb**
  Detonated by the USA in New Mexico, USA.

- **1948 Computer**
  Manchester Mark I, the world's first computer, was built in Manchester, England.

1950-Today

- **1950 Kidney transplant**
  Performed by R.Lawler, USA.

- **1953 DNA**
  The structure of DNA was worked out by F.Crick, England and J.Watson, USA.

- **1954 Nuclear power station**
  Built to generate electricity at Obninsk, USSR.

- **1955 Nuclear submarine**
  USS Nautilus launched in the USA.

- **1955 Hovercraft**
  The principle of the hovercraft was patented by C.Cockerell, England.

- **1957 Satellite**
  Sputnik I, the first satellite, was launched by the Russians.

- **1958 Silicon chip**
  Patented by Texas Instruments, USA.

- **1960 Laser**
  Built by T.Maiman, USA.

- **1960 Oral contraceptive**
  The first comercially-produced oral contraceptive pill was Enovid 10, made by G.D.Searle Drug Co., USA.

- **1961 Manned spaceflight**
  Made by Yuri Gagarin, USSR.

- **1964 Word processor**
  Introduced by IBM, USA.

- **1967 Heart transplant**
  Performed by Christiaan Barnard, South Africa.

- **1971 Space station**
  Salyut 1 launched by the USSR.

Salyut 1 space station ▲

- **1978 Test-tube baby**
  The first test-tube baby was born at Oldham, England.

- **1981 Space Shuttle**
  'Columbia' made its maiden flight in orbit round Earth.

- **1985 Anti-satellite missile**
  Launched by the USA.

A - Z of inventors

Ampère, André-Marie 1775-1836
French scientist. The first person to understand the importance of the relationship between **electricity and magnetism**. He designed the earliest form of **electromagnet**.

Appert, Nicolas-François c.1750-1841
French inventor of **air-tight containers** for preserving food.

Archimedes c.287-212BC
Greek mathematician and philospher. Believed to have invented the **Archimedean screw** used for raising water. He formulated the law of buoyancy, known as 'Archimedes' Principle'.

Aristotle 384-322BC
Greek scientist and philosopher who laid the foundations of **modern scientific study**.

Arkwright, Sir Richard 1732-92
English wigmaker and industrialist who invented a revolutionary **spinning machine**.

Babbage, Charles 1792-1871
English mathematician who designed an **analytical engine**, the ancestor of the modern calculator.

Bacon, Friar Roger c.1220-92
English monk and scientist. First described a **magnifying glass**. He was also one of the first people who believed the Earth was round.

Baekeland, Leo Hendrick 1863-1944
Belgian-American chemist and industrialist. Invented the first heat-proof plastic, called **Bakelite**.

Baird, John Logie 1888-1946
Scottish engineer. Produced the first successful **television pictures**. He also discovered the principle of **fibre optics**.

Banting, Sir Frederick 1891-1941
Canadian scientist who, with Charles Best, developed **insulin**, a drug that helps diabetics.

Barnard, Christiaan 1923-
South African physician. Performed the first successful **heart transplant on a human**.

Becquerel, Antoine Henri 1852-1908
French physicist. Discovered **radioactivity**.

Bell, Alexander Graham 1847-1922
Scottish-American teacher. Invented the **telephone**.

Benz, Karl 1844-1929
German engineer and inventor of the first practical **car powered by an internal combustion engine**.

Berliner, Emile 1851-1929
German-American. Invented the **gramophone disc**.

Bessemer, Sir Henry 1813-98
English engineer. Patented the first successful method of **mass-producing steel**, using the Bessemer converter.

Biro, Ladislao 1900-85
Hungarian-Argentinian artist. Invented the first practical **ballpoint pen**.

Bissell, Melville 1843-89
American inventor of the **carpet sweeper**.

Bohr, Niels 1885-1962
Danish physicist. First proposed a model of the **atom based on quantum physics.**

Boyle, Robert 1627-91
Anglo-Irish physicist and chemist. Made major discoveries about the **properties of gases**. First defined the modern idea of an **element**.

Braille, Louis 1809-52
French teacher. Invented the **Braille reading and writing system** for blind people.

Bramah, Joseph 1748-1814
English cabinet-maker. Designed the first practical **lavatory**, the **'Bramah' lock** and the **hydraulic press**.

Braun, Baron Wernher von 1912-74
German rocket engineer who invented the first true missile, the **V-2**.

Broglie, Louis Victor, Duc de 1892-
French physicist. First suggested that matter, like light, could act as waves, known as **matter waves**.

Brunel, Isambard Kingdom 1806-59
English engineer. Famous for his **ships**, **bridges** and **railways**.

Carlson, Chester 1906-68
American scientist who invented the **photocopying process**.

Carothers, Wallace Hume 1896-1937
American chemist. Invented **nylon**.

Cartwright, Rev Edmund 1743-1823
English clergyman. Invented the **power loom**.

Cayley, Sir George 1773-1857
English designer and builder of the first **man-carrying glider**.

Chadwick, Sir James 1891-1974
English physicist. Discovered the **neutron**.

Chappe, Claude 1763-1805
French inventor of the **semaphore signalling system**.

Cierva, Juan de la 1895-1936
Spanish engineer and inventor of the **autogyro**.

Cockcroft, Sir John 1897-1967
English scientist who, with Ernest Walton, designed the Cockroft-Walton **particle accelerator**.

Cockerell, Sir Christopher 1910-
English engineer and inventor of the **hovercraft**.

Copernicus, Nicolaus 1473-1543
Polish clergyman and astronomer. Formulated the theory that the **Sun is the centre of the solar system**. This contradicted the Ptolemaic system.

Cousteau, Jacques-Yves 1910-
French oceanographer who invented the **aqualung** for underwater diving.

Crick, Francis 1916-
English biophysicist who, with James Watson, discovered the **double-helix structure of DNA**.

Curie, Marie 1867-1934
Polish-French chemist who, with her husband Pierre, discovered the radioactive elements **polonium** and **radium**.

Cuvier, Georges 1769-1832
French zoologist. Devised a system of **classification of animals**.

Daguerre, Louis 1789-1851
French painter and designer. Invented the first practical photographic process, known as the **daguerreotype process**.

Daimler, Gottlieb Wilhelm 1834-1900
German engineer. Designed the first
successful **internal combustion
engine**.

Dalton, John 1766-1844
English chemist. Worked out the
modern **atomic theory of matter**.

Darby, Abraham c.1678-1717
English iron manufacturer. Perfected
the technique for using **coke** to smelt
iron ore. He built the world's first **iron
bridge**.

Darwin, Charles 1809-82
English naturalist and geologist.
Worked out the **theory of evolution by
natural selection**.

Davy, Sir Humphry 1778-1829
English chemist and inventor of the
miner's safety lamp.

Dewar, Sir James 1842-1923
Scottish scientist. Invented the **vacuum
flask**.

Diesel, Rudolf 1858-1913
German engineer. Developed the
diesel engine.

Dunlop, John 1840-1921
Scottish veterinary surgeon who
invented the first **pneumatic tyre**.

Eastman, George 1854-1932
American industrialist who invented
the first **roll film** for cameras and the
Kodak camera.

Edison, Thomas Alva 1847-1931
American inventor. Patented over a
thousand inventions, including the
light bulb and the **phonograph**.

Ehrlich, Paul 1854-1915
German bacteriologist. Developed the
technique of **chemotherapy**.

Einstein, Albert 1879-1955
German physicist. Formulated the
theories of relativity.

Einthoven, Willem 1860-1927
Dutch physician. Invented the
electrocardiograph.

Faraday, Michael 1791-1867
English physicist and chemist.
Invented the **electric motor** and
transformer.

Fermat, Pierre de 1601-65
French mathematician. Regarded as
the founder of the modern **theory of
numbers**.

Fermi, Enrico 1901-54
Italian physicist. Designed and built
the first **nuclear reactor**.

Fleming, Sir Alexander 1881-1955
Scottish bacteriologist. Discovered the
antibiotic **penicillin**.

Ford, Henry 1863-1947
American car manufacturer and the
pioneer of **mass-production**.

Fox Talbot, William Henry 1800-77
English scientist who invented the first
negative-positive process of
photography.

Franklin, Benjamin 1706-90
American politician and scientist.
Invented **bi-focal lenses** and the
lightning conductor.

Fraunhofer, Joseph 1787-1826
German lens maker who was one of the
founders of the science of
spectroscopy.

Galilei, Galileo 1564-1642
Italian mathematician, physicist,
astronomer. Invented the
thermometer.

Gillette, King Camp 1855-1932
American inventor of the **safety razor**.

Goddard, Robert Hutchings 1882-1945
American physicist, regarded as the father of the **space rocket**.

Gutenberg, Johannes c.1397-1468
German printer who designed and built the first European **printing press**.

Hahn, Otto 1879-1968
German chemist who discovered the process of **nuclear fission**.

Hall, Charles Martin 1863-1914
American chemist. Invented the **electrolytic extraction process for aluminium**.

Hargreaves, James c.1722-78
English weaver. Inventor of the **spinning jenny**.

Harvey, William 1578-1657
English physician who discovered the **circulation of the blood**.

Henry, Joseph 1797-1878
American scientist who built the first practical **electromagnet**.

Herschel, Sir Frederick 1738-1822
German-English astronomer and discoverer of the planet Uranus.

Hertz, Heinrich Rudolf 1857-94
German physicist who discovered the existence of **radio waves**.

Hero of Alexandria Ist century AD
Greek-Egyptian engineer and inventor of the first **steam-powered machine**, made up of a rotating metal sphere.

Hippocrates 460-377BC
Greek physician who laid the foundations of **medical diagnosis** based on observation.

Huygens, Christiaan 1629-95
Dutch scientist who made a high-resolution **telescope** and **pendulum clock**.

Jenner, Edward 1749-1823
English doctor. Discovered the **principles of vaccination**.

Jouffroy d'Abbans, Claude, Marquis de c.1751-1832
French soldier and engineer who invented the first practical **steamboat**.

Kepler, Johannes 1571-1630
German astronomer. Worked out the basic laws of the **motion of planets**.

Koch, Robert 1843-1910
German physician who founded the science of **bacteriology**.

Land, Edwin Herbert 1909-
American physicist and inventor of the **polaroid camera**.

Landsteiner, Karl 1868-1943
Austrian physician who discovered the **ABO blood group system**.

Lavoisier, Antoine-Laurent 1743-94
French chemist. Drew up the first system of **chemical nomenclature**.

Leeuwenhoek, Antonie van 1632-1723
Dutch scientist who made microscopes and extremely accurate hand-held **lenses** and was the first person to study bacteria.

Lenoir, Etienne 1822-1900
Belgian engineer. Designed the first, gas fired, **internal combustion engine**.

Linné, Carl 1707-78
Swedish botanist, also known as **Linnaeus**. Developed a **system of classifying plants** that forms the basis of modern classification.

Lippershey, Hans c.1570-c.1619
Dutch spectacle-maker. Invented the **telescope**.

Lister, Joseph, Lord 1827-1912
English surgeon who introduced **antiseptic operations** in hospitals.

Lumière, Auguste and **Louis** 1862-1954 and 1864-1948
French inventors of the cinematographe, the first **motion-picture camera**.

Mach, Ernst 1838-1916
Czechoslovakian physicist. Worked out 'Mach numbers', the ratio of the **speed of sound** and the speed of a body in undisturbed air.

Marconi, Guglielmo 1874-1937
Italian physicist. Invented the first practical system of **wireless telegraphy**.

Maxim, Sir Hiram Stevens 1840-1916
American-English engineer. Invented the first **fully automatic machine gun**.

Maxwell, James Clerk 1831-79
Scottish mathematician who worked out the **field theory of electricity and magnetism**.

McAdam, John Loudon 1756-1836
Scottish engineer. Invented the **macadamized road surface**.

Mendel, Gregor Johann 1822-84
Czechoslovakian monk and botanist who worked out the **theory of inheritance**. Founder of the science of genetics.

Mendeleyev, Dmitry 1834-1907
Russian chemist. Worked out the **periodic table of elements**, now regarded as the backbone of modern chemistry.

Montgolfier, Jacques and **Joseph** 1745-99 and 1740-1810
French inventors of the **hot-air balloon**.

Morse, Samuel 1791-1872
American inventor of the **Morse code**.

Newcomen, Thomas 1663-1729
English inventor of the first practical **steam engine**.

Newton, Sir Isaac 1642-1727
English mathematician, physicist, astronomer and philosopher. Invented the **reflecting telescope** and the **calculus**, discovered the **spectrum** and worked out the three **laws of motion** and the **theory of gravitation**.

Niépce, Joseph-Nicéphore 1765-1833
French scientist who produced the first **fixed photographic image**.

Nipkow, Paul 1860-1940
German scientist. Invented the **nipkow disc**.

Nobel, Alfred 1833-96
Swedish inventor of **dynamite** and the founder of the Nobel Prizes.

Otto, Nikolaus 1832-91
German engineer. Built the first **four-stroke internal-combustion engine**.

Parsons, Sir Charles 1854-1931
Anglo-Irish engineer. Invented the practical **steam turbine**.

Pascal, Blaise 1623-62
French philosopher and mathematician who invented the mechanical **adding machine**. With Fermat, he investigated the theory of probability.

Pasteur, Louis 1822-95
French chemist who established that **germs** were the cause of fermentation and disease. He also invented the process of **pasteurization**.

Perkin, Sir William 1838-1907
English chemist who discovered the first **synthetic dye**, called mauveine, in 1856.

Planck, Max 1858-1947
German physicist who worked out the **quantum theory**, c.1900. This states that energy acts in tiny packages called 'quanta'.

Priestley, Joseph 1733-1804
English scientist and discoverer of
oxygen.

Ptolemaeus, Claudius 2nd century AD
Greek astronomer, geographer and
mathematician, also known as
Ptolemy. Formulated the **Ptolemaic
system of the Universe**.

Röntgen, Wilhelm 1845-1923
German physicist. Discovered **X-rays**.

Rutherford, Sir Ernest 1871-1937
New Zealand scientist. Constructed
the first **nuclear model of the atom** and
also **split the atom**.

Salk, Jonas Edward 1914-
American virologist who developed
the **vaccine for poliomyelitis**.

Senefelder, Aloys 1771-1834
Czechoslovakian inventor of
lithography.

Sholes, Christopher Latham 1819-90
American inventor of the first practical
typewriter.

Sikorsky, Igor Ivanovich 1889-1972
Russian-American engineer who
designed the modern **helicopter**.

Stephenson, George 1781-1848
English mechanic and designer of the
Rocket steam engine, the first
successful steam train.

Swan, Sir Joseph 1828-1914
English scientist. Invented a **light bulb**,
at about the same time as Edison, and a
vacuum pump for light bulbs.

Tesla, Nikola 1856-1943
Yugoslavian-American inventor of the
induction motor, as well as generators,
dynamos and a transformer.

Thomson, Sir Joseph 1856-1940
English physicist. Discovered the
electron.

Torricelli, Evangelista 1608-47
Italian physicist. Invented the
barometer.

Townes, Charles Hard 1915-
American inventor of the **laser**.

Vinci, Leonardo da 1452-1519
Italian artist and scientist who designed
many devices, none of which were
published.

Volta, Count Alessandro 1745-1827
Italian scientist. Invented the voltaic
pile, the first **electric battery**.

Wankel, Felix 1902-
German engineer. Designed the
rotary engine.

Watson, James 1928-
American scientist who, with Francis
Crick, discovered the **double helix
structure of DNA**.

Watt, James 1736-1819
Scottish engineer. Developed the
steam engine. The unit of electricity
called the "watt" is named after him.

Whitney, Eli 1765-1825
American inventor of the **cotton gin**.

Whittle, Sir Frank 1907-
English engineer and inventor of the
jet engine.

Wright, Wilbur and **Orville** 1867-1912
and 1871-1948
American designers of the first
successful **power-driven aeroplane**.

Zworykin, Vladimir Kosma 1889-
Russian-American physicist and
inventor of the **iconoscope**.

Glossary

Alternating current
This is an electric current in which the electrons move backwards and forwards in the wire instead of in one direction.

Constantinople
The former name of Istanbul, capital of Turkey. It was founded by the Roman Emperor Constantine.

Diaphragm
A thin metal plate that vibrates when receiving or producing sound waves. It is used to convert sound signals to electrical signals and vice versa in telephones and microphones.

Ellipse
A closed area in the shape of a flattened circle.

Filament
The thin wire, usually made of tungsten, inside a light bulb. It gives off light when heated by an electric current.

Inflammable
Something which is liable to catch fire.

Isotope
Atoms of an element which have different mass numbers are different isotopes of that element.

Mass
A physical quantity expressing the amount of matter in a body or object.

Matter
The material which makes up something, in particular a physical object.

Molecule
The simplest unit of a chemical compound that can exist. It is made up of two or more atoms held together by chemical bonds.

Probe
A space probe is an unmanned vehicle that obtains scientific information, usually about the atmosphere and surfaces of planets. It sends the information back to Earth as radio messages or television pictures.

Solar eclipse
This occurs when the Moon passes between the Sun and Earth, blocking out the light from the Sun for a short while.

Solar system
The system containing the Sun and the planets held in its gravitational field, including the Earth, Mercury, Venus, Mars, Jupiter, Saturn, Uranus, Neptune and Pluto.

Transducer
An electrical device that sends out and receives sound waves.

Turbofan
A type of engine in which a large fan sucks in air and forces it backwards. This increases the power of the engine.

Vacuum
A closed space that contains nothing. Everything, including air, has been removed. The pressure outside the vacuum is greater than inside.

Index

Index of illustrations

E

Edison's kinetoscope, 58
        light bulb, 62
        phonograph, 54
Egyptian balance, 70
        right-angle, 48
        sailing boat, 32
        shadow clock, 68
        water clock, 68
Einstein's theories of
relativity, 81
Electric drill, 19
Electric fire, 63
Electric sewing machine, 61
Electric telegraph, 108
Electric typewriter, 45
Electrocardiograph, 85
Electromagnet, 7
Electron microscope, 77
Electrostatic accelerator, 15
Endoscope, 84
English galleon, 93
USS Enterprise, 93
Ether flask, 86
European numbers, 48

F

Fan heater, 63
Faraday's electric motor, 7
Fast breeder reactor, 17
Fermi's nuclear reactor, 17
Fibre optic cable, 52
Film soundtrack, 58
Fleming's diode valve, 110
Float glass, 25
Fluorescent light, 62
'Flyer 1', 34
Food mixer, 64
Food processor, 64
Ford Model T, 30
Fountain pen, 45
Four-colour printing, 46
Four-stroke engine, 9
Fractioning column, 5

G

Galilei's thermoscope, 74
Galleon, 32
Gas drilling rig, 5
Gas mask, 94
Geothermal power station, 13
Gesner's pencil, 103
GE Vulcan machine gun, 91
Giant catapult, 90

Giffard airship, 34
'Great Britain', 32
Greek armour, 94
Greek trireme, 93
Groin vault, 24
Guericke's air pump, 104
Gyroscope, 20

H

Halogenheat, 65
Harrier 'Jump Jet', 35
Harrison's marine
chronometer, 106
Heart-lung machine, 87
Heart transplant, 87
Hero's steam engine, 100
Hertz's radio equipment, 53
Hieroglyphs, 44
Hindu numerals, 48
'Holland VI', 33
Holography, 57
Home computer, 50
Hooke's microscope, 77
Hot rocks station, 13
Hover mower, 66
Hurricane Gladys, 75
Hydraulic press, 20
Hydroelectric power station, 10
Hydrofoils, 75
Hygrothermograph, 75

I

Impulse turbine, 10
Interferometers, 79
Ionosphere, 53
Iron lung, 87

J

Jacquard loom, 60
Japanese armour, 94
Jet engine, 9
Jethro Tull's seed drill, 40
Junkers J1, 34

K

Kepler's laws of the motion
of planets, 80
Kidney machine, 87

L

Laënnec's stethoscope, 107
LAGEOS, 71
Larrey's ambulance, 107
Laser, 21
Laser lightwaves, 21
Lenoir's gas engine, 8
Letterpress printing, 46
Light bulb, 62
Lightning conductor, 6
Light spectrum, 79
Lightwaves, 21
Lilienthal's hang glider, 34
Linnaeus' system of
classification, 89
Lippershey-Galilei
telescope, 104
Liquid fertilizer spraying, 42
Lister's carbolic spray, 83
Lithography, 47
Locks (water gates), 27
Lodestone, 99
Logarithms, 48
Longbow, 102
Long radio wave, 53
Loom, 60
Luna 9, 37
Lydian coin, 67

M

M-107 cannon, 91
Macmillan's bicycle, 29
Magnifying glass, 76
Magnetic field, 7
Magnets, 7
Mail-coach and horses, 67
Mariner 4, 38
Matchlock mechanism, 91
Maudslay's metal-working
lathe, 18
Mechanical clock, 68
Medium radio wave, 53
Mendel's experiment on
sweet peas, 89
Mendeleyev's periodic
table, 88
Mercator's map, 72
Michaux motorcycle, 29
Microprocessor chip, 50
Microscope, 77
Milking machines, 41
Miner's helmet, 4
Moderator atoms, 16
Modern car and its parts, 31
Modern helicopter, 35
Modern military clothing, 94
Monotype typesetting, 47
Montgolfier balloon, 34